The Al Kluis
Farmer's Almanac

Kluis Publishing -- Wayzata

Kluis Publishing, LLC
901 Twelve Oaks Center Drive, Suite 907, Wayzata, MN 55391
Visit our website at www.AlKluis.com

For information about special discounts for bulk purchases, write to: info@kluispublishing.com.

First edition

2 4 6 8 10 9 7 5 3 1

ISBN: 978-0-9830382-3-8

Book design by Michael Iedema

Cartoons copyright 2013 by Bill Lee

Poems copyright 2013 by Darren Sardelli

Printed in the United States of America

Information Sources:
- Crop data and projections, unless otherwise specified, are from USDA (U.S. Department of Agriculture) reports.
- Seasonal odds are calculated by Al Kluis using CBOT (Chicago Board of Trade) price data.
- Prior-year closing prices are from the CBOT.

Greetings!

Wow. It has been a year of records. The drought of 2012 started in South America and also showed up in the Corn Belt. The result was record high corn and soybean prices. Livestock producers who were not hedged had a challenging year.

This was a challenging year to be in the commodity business overall. The low came in late June, then grain prices went higher in a very orderly rally that peaked in late August to early September. Meanwhile, the European Central Bank kept printing money, as did the Federal Reserve Board in the US, and farm land was again a top performer. Still, farmers who had a trendline or better crop--and who were disciplined sellers using all of their marketing alternatives--had another great year. For those who were in the middle of the drought, it was a crop insurance year.

Put it all together, and we are set up for another volatile year in 2013.

What did I learn from 2012? I learned that decisions about crop insurance were the most important risk management decisions many farmers made in 2012. The farmers who stuck with my three-step risk management plan had a good year, even if their crop was a total disaster. Here, in a nutshell, is the three-step plan: (1) Get the right crop insurance plan bought, (2) Place scale-up hedges against your insured bushels, and (3) Cover the rest of them with puts. After two years of counter-seasonal moves, the odds are that we will have a more typical pricing year in 2013 with a harvest low.

The majority of my 2012 Fearless Forecasts worked. The forecast for hog farming to be profitable in 2012 was really a bust, as low hog prices and high feed prices put a squeeze on margins. And one of these years, I will stop forecasting crude oil prices, which I have yet to get right. Prices went higher than I thought possible with the chaos in the Middle East. But I like to watch and chart crude oil prices. It helps me think through the right buy recommendations for the farmers who follow my advice.

In the pages of this 2013 Almanac, you will see my Fearless Forecasts for 2013. My early prognostication is that the majority of my Fearless Forecasts will be right again.

I hope you enjoy our 2013 Farmers Almanac, and use it to make better, more profitable decisions.

Sincerely,

Al Kluis

Al Kluis
Wayzata, Minnesota

ways to use this Almanac in 2013

When I first decided to write an all-new Almanac for today's farmers, I wanted it to be useful, something you can use to learn and make better decisions. (Well, with a few things thrown in for fun.) You work hard, and here are some tips to make your Almanac work just as hard for you:

 Use it. Read it from front to back. The re-read it each month. Write on it. Make your own fearless forecasts, or take odds on which ones of mine will work. Discuss the Almanac and the information in it with your marketing team. Keep one copy of the Almanac in your grain truck  and one in the office. If you have a long line as you are hauling corn, at least you can learn while you wait.

 Take time each week to look at the data for the week. Write on it, and keep track of how many gallons of gas you can buy with a bushel of corn or how many gallons of diesel you can buy with a bushel of soybeans. My ideas are not useful for day trading, but I do put my 37 years of knowledge into this book. By reading this and writing down your own thoughts and observations, you can learn all year long. I believe the long-term trends and projections I explain will work most of the time.

 Share the Almanac with someone new to the farm. Have your son or son-in-law or daughter keep up the charts or the cash-o-meter. This is designed as an educational tool, and by sharing it you can both learn more. Watch how often the five year seasonal odds works, and note when it does not. Even  today, there are very few classes on grain marketing even at the college level.

Study the Almanac on the major USDA crop report days. Put your own estimates in the day ahead using a red pen, then plug in the numbers from the USDA using a blue pen. How close were you? Compare the numbers to what the trade had expected and what they were last year. This all helps you learn more and give you a long term perspective. A good marketer has a sense of history.

 It is OK to make copies of the long-term charts and the USDA tables. Look at the trends the USDA is showing and then compare it to your farm. Are you planting more corn when everyone else is? If you are a crop farmer it is important to stay aware of livestock numbers. The tools are all here. You need to take time to use them.

 Write to me or e-mail me at info@kluispublishing.com with ideas on how I can improve the Almanac, or to tell me which part of the Almanac is the most useful for you. Tell me (ahead of time!), which of my Fearless Forecasts will be wrong and why. Most important, let me know how this Almanac has helped you.

Good luck in 2013.

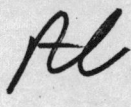

30-5

5-Year Seasonal Odds this Week		
	change	reliability
Corn	-1.0¢ ⇩	60%
Soybeans	+6.9¢ ⇧	40%

New Crop This Week

	Hi	Lo	Close	12 Close
December Corn				$5.75
November Soybeans				$11.91

Old Crop This Week

	Hi	Lo	Close	12 Close
March Corn				$6.44
January Soybeans				$11.90

Sunday — 30

Monday — 31

Tuesday — 1
- CBOT CLOSED: New Year's Day

Wednesday — 2

Thursday — 3

Friday — 4

Saturday — 5

My 2013 Beans: _____ acres X _____ bu/acre X $ _____ per bushel + $ _____
My 2013 Corn: _____ acres X _____ bu/acre X $ _____ per bushel + $ _____

Al's Saturday Cash-o-Meter

This is why you work so hard! (Well, at least one of the reasons.)

Every week, calculate the value of this year's crop. As you plan, and as the season progresses, this number can--and will--change dramatically. The acres you plant, then harvest, can change until the day you fire up the combine. The yield per acre changes throughout the growing season, then can still give you some surprises when you roll the grain wagon across the scale. And the price per bushel... well, you know how much that can change!

Fill in the Cash-o-Meter every Saturday and you will see how dramatically this number changes in the course of the year. It is a good reminder of how important it is to have a marketing plan, and to use it.

Al's Fearless Scorecard

Al says: "How did I do with my 2012 Fearless Forecasts?"

Date In 2012	Forecast	Result
January 15 to 21	"December 2012 Corn futures will drop below the December 2011 Corn harvest low by October of 2012."	Yes. December 2012 Corn dropped down to a June 2012 low at $5.51 that was 21 cents lower than the low of December 2011 corn in October of 2011.
January 22 to 28	"Soybeans will rally above $15.00 by the end of August 2012."	Yes. Soybean futures went up to new all time highs, trading as high as $17.80 by August of 2012.
February 5 to 11	"Crude oil will drop below $70.00 per barrel by the third quarter of 2012."	No. The low came in on June 28, 2012 at $78.28. About $8 higher than my projection, but the low occurring in June was right.
February 12 to 18	"Hog farming will be very profitable in 2012."	No. Hog farmers lost a lot of equity in the second and third quarters of 2012 as feed prices went up and hog prices went down.
February 26 to March 3	"Stocks will be more profitable than bonds in 2012."	Yes. At the time I am writing this the US stock market is up about 12% and bonds are up 3%.
March 4 to 10	"US farmers will plant over 95 million acres of corn in 2012."	Yes. US farmers planted 96.4 million acres of corn in 2012, the most in over 40 years.
April 8 to 14	"Wheat prices will rebound by the third quarter of 2012."	Yes. In January 2012, wheat prices were $6.75 to $7.25 per bushel, and by the third quarter of 2012 wheat prices were trading at $8.75 to $9.50 per bushel

Watch for the 2013 "Fearless Forecasts" in upcoming pages.

Date In 2012	Forecast	Result
April 22 to 28	"US and global stock markets will rally back by the third quarter of 2012."	Yes. By September of 2012 US stock prices were making new four year highs. Global stock markets were following the strength in the US market.
April 29 to May 5	"Land prices will continue to outperform the stock market in 2012."	Yes. Corn Belt land values were up another 18 to 20%, beating the 12% gain in the US stock market.
May 6 to 12	"Global ending stocks of soybeans will drop to 20 days usage in 2012."	No. Global ending stocks dropped down to about 20% of usage or about 70 days.
May 20 to 26	"Wheat stocks will fall in the third quarter of 2012."	Yes. Because of smaller crops in Europe, Russia, and the Ukraine, world wheat stocks dropped down to 177 million metric tons, which is down 21 million metric tons from last year.
June 3 to June 9	"China's growth rate will drop to just 8%."	Yes. The most recent data projects the growth rate at 7.4% with the current estimate of 8% growth for the year.

Watch for the 2013 "Fearless Forecasts" in upcoming pages.

2012 "New Year Price Indicator"

		High			Low			Close		
		Week 1	Week 2	Change	Week 1	Week 2	Change	Week 1	Week 2	Change
Corn										
March 12 (old crop)		6.64	6.63	-	639	5.98	-	6.43	6.00	-
Dec 12 (new crop)		5.97	5.89	-	5.75	5.53	-	5.75	5.55	-
Soybeans										
March 12 (old crop)		12.44	12.34	-	11.94	11.51	-	11.96	11.58	-
Nov 12 (new crop)		12.31	12.24	-	11.88	11.63	-	11.91	11.70	-

How it works: I use the "New Year Price Indicator" to get a hint about the first quarter of the year. The key is the price action in the first two weeks of January: Do the weekly highs, lows, and closes go up? Or do they go down? That sets the tone for the quarter.

How it worked in 2012: The New Year Price Indicator suggested a lower price for corn and soybeans. Corn sold off into February but bounced back by the end of March. Soybeans rallied from the mid-January low right through the end of March as dry weather rallied global soybean prices.

2013 "New Year Price Indicator"

	High			Low			Close		
	Week 1	Week 2	Change	Week 1	Week 2	Change	Week 1	Week 2	Change
Corn									
March 13 (old crop)									
Dec 13 (new crop)									
Soybeans									
March 13 (old crop)									
Nov 13 (new crop)									

6-12

5-Year Seasonal Odds this Week		
	change	reliability
Corn	**-12.0¢** ⬇	**60%**
Soybeans	**-12.0¢** ⬇	**80%**

New Crop This Week

	Hi	Lo	Close	12 Close
December Corn				$5.55
November Soybeans				$11.70

Old Crop This Week

	Hi	Lo	Close	12 Close
March Corn				$5.99
January Soybeans				$11.58

Sunday   6

Monday   7

Tuesday   8
- Al's "Second Tuesday" Webinar

Wednesday   9

Thursday   10

Friday   11
- USDA report: Grain Stocks, WASDE, Crop Production, Crop Production Annual

Saturday   12

My 2013 Beans: _____ acres X _____ bu/acre X $ _____ per bushel + $ _____
My 2013 Corn: _____ acres X _____ bu/acre X $ _____ per bushel + $ _____

Dutch Tip

"Bad Save": Storing poor quality corn never pays off. It can be an expensive mistake.

	US Corn Supply and Use *(millions of bushels)*			
	Crop Year			
Item	2010	2011	2012 Estimated (as of Sept 2012)	2013 Projected (as of Jan 11, 2013)
Area planted (mil. ac.)	*88.2*	*91.9*	*96.4*	
Area harvested (mil. ac.)	*81.4*	*84.0*	*87.4*	
Yield per acre (bu)	*152.8*	*147.2*	*122.8*	
Beginning stocks	1,708	1,128	1,181	
Production	12,447	12,358	10,727	
Imports	28	25	75	
Total Supply	**14,182**	**13,511**	**11,983**	
Feed and residual	4,793	4,400	4,150	
Food, seed & industrial*	6,428	6,390	5,850	
Exports	1,834	1,540	1,250	
Total Use	**13,055**	**12,330**	**11,250**	
ENDING STOCKS	**1,128**	**1,181**	**733**	

* This includes:

Corn used for ethanol	*5,021*	*5,000*	*4,500*	

Marketing Check List

☐ Review your 2012 grain sales. List these things: Your highest sale, your lowest sale, basis on the date you sold, and average selling price.

	Crop Year			
Item	**2010**	**2011**	**2012** *Estimated* *(as of Sept 2012)*	**2013** *Projected* *(as of Jan 11, 2013)*
Area planted (mil. ac.)	*77.4*	*75.0*	*76.1*	
Area harvested (mil. ac.)	*76.6*	*73.6*	*74.6*	
Yield per acre (bu)	*43.5*	*41.5*	*35.3*	
Beginning stocks	151	215	130	
Production	3,329	3,056	2,634	
Imports	14	16	20	
Total Supply	**3,495**	**3,287**	**2,785**	
Crushings	1,648	1,705	1,500	
Exports	1,501	1,360	1,055	
Seed	87	88	89	
Residual	44	3	25	
Total Use	**3,280**	**3,157**	**2,670**	
ENDING STOCKS	**215**	**130**	**115**	

US Soybean Supply and Use
(in millions of bushels)

USDA, Division of Motion Pictures, 1941

13-19

5-Year Seasonal Odds this Week		
	change	reliability
Corn	+2.7¢ ⇧	40%
Soybeans	-4.1¢ ⇩	60%

New Crop This Week

	Hi	Lo	Close	12 Close
December Corn				$5.52
November Soybeans				$11.84

Old Crop This Week

	Hi	Lo	Close	12 Close
March Corn				$6.12
March Soybeans				$11.87

Sunday
13

Monday
14
- Last Trading Day: Jan Soybeans

Tuesday
15

Wednesday
16
- Last Delivery Day: Jan Soybeans

Thursday
17

Friday
18

Saturday
19

My 2013 Beans: _____ acres X _____ bu/acre X $ _____ per bushel + $ _____
My 2013 Corn: _____ acres X _____ bu/acre X $ _____ per bushel + $ _____

 # Al's Fearless Forecast

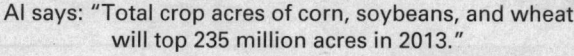

Al says: "Total crop acres of corn, soybeans, and wheat will top 235 million acres in 2013."

	Date	2010	2011	2012 (as of Sept 2012)	2013 (as of Jan 11, 2013)
Corn	Mar 1	7,693,787	6,523,228	6,023,356	
	Jun 1	4,310,071	3,669,838	3,148,540	
	Sep 1	1,707,787	1,127,645	988,403	
	Dec 1	10,056,769	9,647,466		
Soybeans	Mar 1	1,270,068	1,248,800	1,374,488	
	Jun 1	571,123	619,283	667,475	
	Sep 1	150,885	215,013	169,417	
	Dec 1	2,278,084	2,369,885		

What's left?

*US **grain stocks** stored on and off the farm (in thousands of bushels)*

Marketing Check List

❏ Build your 2013 old crop and 2013 new crop continuation charts and assign someone on your team to keep them up.

20-26

5-Year Seasonal Odds this Week		
	change	reliability
Corn	-1.0¢ ⇩	60%
Soybeans	-0.9¢ ⇩	60%

New Crop This Week

	Hi	Lo	Close	12 Close
December Corn				$5.71
November Soybeans				$12.22

Old Crop This Week

	Hi	Lo	Close	12 Close
March Corn				$6.42
March Soybeans				$12.19

Sunday — 20

Monday — 21
- CBOT CLOSED: Martin Luther King Day

Tuesday — 22

Wednesday — 23

Thursday — 24

Friday — 25
- USDA report: Cattle On Feed

Saturday — 26

My 2013 Beans: _____ acres X _____ bu/acre X $ _____ per bushel + $ _____
My 2013 Corn: _____ acres X _____ bu/acre X $ _____ per bushel + $ _____

US Soybean Supply and Use *(in millions of bushels)*				
	Crop Year			
Item	**2010**	**2011**	**2012** *Estimated (as of Sept 2012)*	**2013** *Projected (as of Jan 11, 2013)*
Area planted (mil. ac.)	*77.4*	*75.0*	*76.1*	
Area harvested (mil. ac.)	*76.6*	*73.6*	*74.6*	
Yield per acre (bu)	*43.5*	*41.5*	*35.3*	
Beginning stocks	151	215	130	
Production	3,329	3,056	2,634	
Imports	14	16	20	
Total Supply	**3,495**	**3,287**	**2,785**	
Crushings	1,648	1,705	1,500	
Exports	1,501	1,360	1,055	
Seed	87	88	89	
Residual	44	3	25	
Total Use	**3,280**	**3,157**	**2,670**	
ENDING STOCKS	**215**	**130**	**115**	

Marketing Check List

❏ Record the high, low and close for the first 10 days of the marketing year. Do this for old crop and new crop soybeans.

Your 2013 Marketing Plan

It looks like 2013 will be a challenging year. There are five factors you need to evaluate when you put together your 2013 marketing plan.

1. Be able to lock in most of your 2013 inputs for less than your 2012 input costs. The big variable is land rents, which for many farmers will not be set until March 2013.

2. Be aware that the global financial uncertainty continues. The Euro debt problems have been going on now for three years. This has rallied the dollar and slowed down global economic growth. This will limit how far grain prices can rally, and it could result in much lower prices in 2013, especially if China goes into a hard landing.

3. You won't know your revenue insurance price guarantee until March 1, 2013. So, you'll have a lot of risk until you know what the price level of that guarantee is.

4. With the current 2013 soybean corn ratio at 2.24:1, corn looks like it will make more money in 2013 than soybeans.

5. The 2013 corn and soybean futures are trading at a discount to the 2012 crop futures. You should be able to place some hedges at higher price levels during a North or South American weather scare in 2013.

Some Action Steps

Last fall, I suggested farmers lock in their biggest inputs. This meant trying to get their fall fertilizer prices locked in, and checking on the cost and availability of seed for 2013. Hopefully you have done those things. In addition, if possible, I suggested farmers get their land agreements signed for 2013 before the 2012 harvest. If you weren't able to do those things last year, make a note to yourself to do them later this year for the 2014 season.

Once your major inputs are locked in, you will be more comfortable getting some 2013 hedges in place. You cannot control the global uncertainty or know (before March 1) what your RP guarantee will be, so don't waste any nervous energy worrying about those two factors.

Watch the price of December 2013 crop corn. If corn prices rally and the ratio gets down to 2:1, shift more acres to corn and get a large part of those additional bushels sold ahead. If you're able to get your inputs locked in and December 2013 corn and November 2013 soybean futures rally up to 20¢ off the contract highs, get 10% to 20% of your 2013 crop hedged ahead.

Every farm is different in how it approaches marketing and how it evaluates its critical factors. No matter what your results were in 2012, you need to evaluate your alternatives for 2013, make a plan, and act on it.

US Cattle and Calves: 2012 Inventory
as of Jan. 1, 2012

State	Number (1000 head)	As % of 2011
AL	1,210	98
AK	13	96
AZ	920	106
AR	1,670	97
CA	5,350	103
CO	2,750	104
CT	49	100
DE	19	106
FL	1,710	105
GA	1,020	99
HI	140	99
ID	2,220	100
IL	1,070	97
IN	860	101
IA	3,900	100
KS	6,100	97
KY	2,150	98
LA	790	100
ME	86	96
MD	200	103
MA	41	102
MI	1,110	102
MN	2,360	99
MS	950	106
MO	3,900	99
MT	2,500	100
NE	6,450	104
NV	470	102
NH	35	103
NJ	31	97
NM	1,390	90
NY	1,410	101
NC	810	101
ND	1,690	99
OH	1,280	104
OK	4,500	88
OR	1,300	98
PA	1,610	100
RI	4.5	92
SC	370	96
SD	3,650	99
TN	1,970	99
TX	11,900	89
UT	800	100
VT	260	96
VA	1,490	97
WA	1,110	102
WV	390	105

27-2

5-Year Seasonal Odds this Week		
	change	reliability
Corn	+7.6¢ ⇧	60%
Soybeans	+24.0¢ ⇧	80%

New Crop This Week

	Hi	Lo	Close	12 Close
December Corn				$5.82
November Soybeans				$12.37

Old Crop This Week

	Hi	Lo	Close	12 Close
March Corn				$6.42
March Soybeans				$12.33

Sunday   27

Monday   28

Tuesday   29
• Ag Connect Expo 2013, Kansas City, Missouri

Wednesday   30
• Ag Connect Expo 2013, Kansas City, Missouri

Thursday   31
• Ag Connect Expo 2013, Kansas City, Missouri
• Al's "Last Thursday" Webinar

	Close Today	Close Last Month	Difference
Corn			
Soybeans			

Friday   1
• USDA report: Cattle

Saturday   2

My 2013 Beans: _____ acres X _____ bu/acre X $ _____ per bushel + $ _____
My 2013 Corn: _____ acres X _____ bu/acre X $ _____ per bushel + $ _____

January 1 US Cattle Inventory
(1,000 head)

	2011	2012	2012 as % of 2011	Feb 1, 2013	2013 as % of 2012
Cattle and Calves	100,000	97,800	98		
Cows and Heifers that have calved	40,600	39,700	98		
Heifers > 500 pounds	16,000	15,700	98		
Steers > 500 pounds	14,200	14,000	99		
Cattle on Feed	12,200	12,300	98		
Calf Crop (from prior year)	35,500	35,313.2	99		

USDA, *Division of Motion Pictures, 1934*

Marketing Check List
☐ Check the carrying charges that the market is paying you to store corn and soybeans. Is it enough to justify holding cash grain?

12 Big Words o' Weather
From DT, Al's Favorite Weather Guy

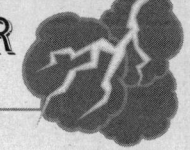

#1. Jet Stream

The jet stream is a global wind "belt" that encircles the middle latitudes. It moves in a pattern made of a series of waves (called "troughs") and hills (called "ridges"). Generally, when a meteorologist refers to the jet stream, we are referring to the middle levels of the atmosphere (500 mb or higher) or the higher levels of 300 or 200 mb. However, the low levels of the atmosphere also have a form of jet streams, though they are not nearly as strong. For the most part it is the high level jet stream that drives weather systems across the planet.

Embedded within this global wind belt are 'jet streaks'. A jet streak is a segment of the jet stream that has significantly stronger winds within the jet stream. Jet streaks (sometimes also called 'jet max') are caused by temperature differences between different levels of the atmosphere. As a result, the jet streaks are often far more intense in the cool season (from October to March) because the difference in temperature between the polar regions and tropical regions is largest then.

Over the past several years, research has showed jet streams can be pulled into low pressure systems, which can cause explosive development of monster systems.

3-9

5-Year Seasonal Odds this Week		
	change	reliability
Corn	+3.6¢ ⇧	60%
Soybeans	-0.1¢ ⇩	60%

New Crop This Week

	Hi	Lo	Close	12 Close
December Corn				$5.60
November Soybeans				$12.39

Old Crop This Week

	Hi	Lo	Close	12 Close
March Corn				$6.32
March Soybeans				$12.29

Sunday 3

Monday 4

Tuesday 5
- LOW: Nov 2013 Soybeans, $9.19 on Feb 5, 2010

Wednesday 6

Thursday 7

Friday 8
- USDA reports: WASDE, Crop Production

Saturday 9

My 2013 Beans: _____ acres X _____ bu/acre X $ _____ per bushel + $ _____
My 2013 Corn: _____ acres X _____ bu/acre X $ _____ per bushel + $ _____

In the USDA Crop Production Report this Month:
Papayas, Citrus Fruits, Sugarcane

About those millibars...

Meterologists don't say "winds at 5,000 feet"... they say "winds at 850 mb". In other words, they refer to the pressure at a given height, not the height itself.

Also called....	Pressure in millibars (mb)	Approx. height (in feet)	Approx, temp (deg. F)
"Sea level"	1015 mb	0 feet	59 F
	1000 mb	300 ft	59 F
Beginning of "Upper Air" level	850 mb	5000 ft	41 F
	700 mb	10,000 ft	23 F
"Steering level"	500 mb	18,000 ft	-4 F
"Jet stream level"	300 mb	30,000 ft	-49 F
"Top of the atmosphere"	200 mb	40,000 ft	-67 F
	100 mb	53,000 ft	-69 F

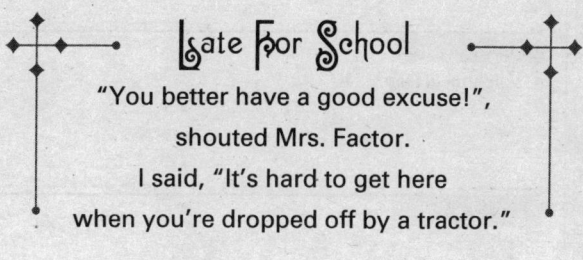

Late For School

"You better have a good excuse!",

shouted Mrs. Factor.

I said, "It's hard to get here

when you're dropped off by a tractor."

Marketing Check List

❑ Make sure you and your marketing team have signed up for at least three marketing seminars or webinars you will attend in the next three months. Sign up for Al's "Second Tuesday" Profit Check webinar on Tuesday night February 12.

February

10-16

5-Year Seasonal Odds this Week		
	change	reliability
Corn	+1.2¢ ⇧	60%
Soybeans	-11.0¢ ⇩	40%

New Crop This Week

	Hi	Lo	Close	12 Close
December Corn				$5.68
Movember Soybeans				$12.62

Old Crop This Week

	Hi	Lo	Close	12 Close
March Corn				$6.42
March Soybeans				$12.68

Sunday — 10

Monday — 11

Tuesday — 12
- Al's "Second Tuesday" Webinar

Wednesday — 13

Thursday — 14
- Valentine's Day

Friday — 15

Saturday — 16

My 2013 Beans: _____ acres X _____ bu/acre X $ _____ per bushel + $ _____
My 2013 Corn: _____ acres X _____ bu/acre X $ _____ per bushel + $ _____

Al-ism

"Write down your marketing plan and you'll have more room in your head for happy thoughts."

Dutch Tip

"Feed 'Em Right": Stock up for your feed needs by buying your soybean meal and corn in late February. You'll save a lot of money by June.

Marketing Check List

☐ Begin checking on spring and fall fuel. Also check on the fall price of LP.

From Al's Mailbox

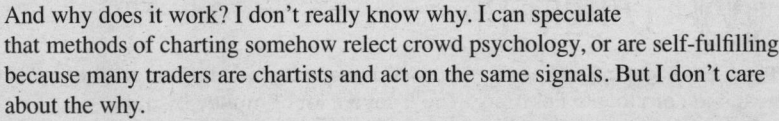

Dear Al,

How can you believe in charts? Do they really work? Why?

Sincerely,
Dedicated Farmer
Missouri, the Show-Me State

Dear Reader:

I started out a skeptic about whether charts could work. I had a degree in Ag Economics from the University of Minnesota, where we studied grain fundamentals. By that I mean we studied supply, demand and the projected ending stocks numbers. That is how I thought you could forecast prices--not by looking at lines on a graph.

In the summer of 1974 I went to my first grain outlook seminar. One of the speakers was a very successful trader from Chicago who explained the basics of chart analysis. He used the charts and several chart analysis tools to make some very accurate price forecasts. I started trading with him and made money using charts.

It is a whole other level of education when you are trading the markets with your own money. You need to learn fast, or you lose fast. I did not make money on every trade, and what I use now is different than my initial approach.

And why does it work? I don't really know why. I can speculate that methods of charting somehow relect crowd psychology, or are self-fulfilling because many traders are chartists and act on the same signals. But I don't care about the why.

What I know after 38 years is that it works. Here are three chart analysis methods my mentor taught me that still work:

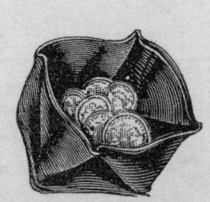

The first of these is the study of price. This is the analysis of trends. In any market you are either in an up trend, down trend or sideways trend. Charts let you see what the trend is and--most importantly--when the trend changes.

Second, the study of motion. This is the study that lets you evaluate when markets are overbought or oversold. I use a combination of two studies: the relative strength index and oscillators. These studies let me see when prices are overbought and likely to top out, and oversold and likely to bottom.

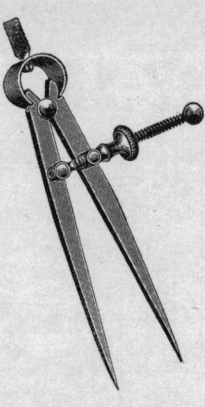

Third and most important is the study of time. This is looking at long-term and short-term cycles, seasonal patterns, and historic high and low data. I mark up my charts as far as 12 to 18 months out with critical change-of-trend weeks. The projections do not always work, but they do alert me to critical time periods when I can anticipate a high or a low.

After 38 years of trading, I now look at the grain fundamentals very differently than when I began. When the fundamentals are the most bullish, prices are near a top. When the fundamentals are the most bearish, you are usually near a low. With that awareness of fundamentals, and by applying all three methods of chart analysis, I have been able to make some money and help my customers make better decisions.

Your dedicated chartist,

Al Kluis

Al Kluis

17-23

5-Year Seasonal Odds this Week		
	change	reliability
Corn	+8.7¢ ⇧	80%
Soybeans	+26.0¢ ⇧	80%

New Crop This Week

	Hi	Lo	Close	12 Close
December Corn				$5.58
November Soybeans				$12.71

Old Crop This Week

	Hi	Lo	Close	12 Close
March Corn				$6.41
March Soybeans				$12.79

Sunday — 17

Monday — 18
- CBOT CLOSED: President's Day

Tuesday — 19

Wednesday — 20

Thursday — 21

Friday — 22
- USDA report: Cattle On Feed

Saturday — 23

My 2013 Beans: _____ acres X _____ bu/acre X $ _____ per bushel + $ _____
My 2013 Corn: _____ acres X _____ bu/acre X $ _____ per bushel + $ _____

Al's Fearless Forecast

Al says: "Live cattle prices will go to new all time highs by August of 2013."

Fighting the Corn Borer with Machinery [1] (Bureau of Agricultural Engineering). 1 reel - 930 feet.

This picture portrays various methods of using machinery to control the corn borer. Of interest to farmers and county agents.

Old Jake Wakes Up (Bureau of Entomology and Plant Quarantine). 1 reel - 978 feet.

A short corn-borer picture designed to awaken interest in the spring clean-up campaign and to supply comedy for corn-borer meetings Includes trick photography, nightmare scenes showing "Old Jake" pursued by fierce corn borers, 10 feet long.

USDA, Division of Motion Pictures, 1934

Marketing Check List

☐ Use a mid-month rally to increase cash corn and soybean sales.

12 Big Words o' Weather
From DT, Al's Favorite Weather Guy

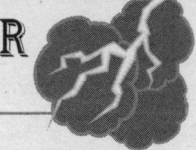

#2. Troughs and Ridges

Although there is a lot of attention given to the surface weather map, it is the position of troughs and ridges in the jet stream that determines where low pressure areas are likely to track. The troughs and ridges also show which areas are likely to stay (or turn) dry.

When meteorologists talk about "scale weather patterns", they are referring to the BIG troughs, which tend to run in a generally north-to-south direction for several hundred-- or even a few thousand--miles.

Likewise, when we refer to the "ridge" position on the large scale, we are referring to a mountain of air that has built up over a particular region. This mountainous ridge will provide a certain kind of weather, usually dry and often warm.

Understanding how troughs and ridges move is very important if you want to understand what kind of pattern will last beyond three or four days. For example: A cold front moving through the Midwest in June may bring showers and thunderstorms with it. But if the pattern supports a ridge over the Midwest, then any sort of showers and thunderstorms associated with this cold front would be generally weak and less significant. And there probably would be no other rain events coming in behind it.

On the other hand, if the cold front is supported by a large or deep trough, then not only is it far more likely to bring severe weather and heavy rain, but there may also be a significant temperature change. If the trough stays over the area, there could be additional storms later in the forecast period.

24-2

5-Year Seasonal Odds this Week		
	change	reliability
Corn	+1.0¢ ⇧	60%
Soybeans	-20.0¢ ⇩	40%

New Crop This Week

	Hi	Lo	Close	12 Close
December Corn				$5.70
November Soybeans				$12.98

Old Crop This Week

	Hi	Lo	Close	12 Close
March Corn				$6.59
March Soybeans				$13.28

Sunday — 24

Monday — 25

Tuesday — 26

Wednesday — 27

Thursday — 28
- Commodity Classic, Kissimmee, Florida
- Al's "Last Thursday" Webinar

	Close Today	Close Last Month	Difference
Corn			
Soybeans			

Friday — 1
- Commodity Classic, Kissimmee, Florida

Saturday — 2
- Commodity Classic, Kissimmee, Florida

My 2013 Beans: _____ acres X _____ bu/acre X $ _____ per bushel + $ _____
My 2013 Corn: _____ acres X _____ bu/acre X $ _____ per bushel + $ _____

Crop Insurance Alert

For most crop insurance products, the industry uses the average price of December Corn futures and November Soy futures during the month of February. It is worth noting that average at the end of February.

Record it here:
December Corn average closing price during
Feb. 2013: _____
November Soybeans average closing price during
February 2013: _____

> **Trees To Tame the Wind** (1 reel, 16 mm. and 35 mm., sound, released
>   1940).
> A narrative story of the planting of field windbreaks and shelterbelts in the
>   prairie States, where farmers are cooperating with the United States Forest
>   Service in one of the most unusual tree-planting programs ever undertaken,
>   in an effort to reclaim land ruined by wind and drought. Trees planted
>   4 or 5 years ago are high enough to indicate their value in the protection of
>   crops and soil. Study guide available on request.

USDA, *Division of Motion Pictures, 1941*

Marketing Check List

☐ This is a key week for livestock feeders and ethanol
   shareholders to buy corn and soybean meal. For grain
   farmers this is a key week to buy calls.

3-9

5-Year Seasonal Odds this Week		
	change	reliability
Corn	-8.7¢ ⇩	60%
Soybeans	-17.0¢ ⇩	60%

New Crop This Week

	Hi	Lo	Close	12 Close
December Corn				$5.63
November Soybeans				$13.05

Old Crop This Week

	Hi	Lo	Close	12 Close
March Corn				$6.54
March Soybeans				$13.32

Sunday — 3
- LOW: Nov 2011 Soybeans, $8.23 on Mar 3, 2009

Monday — 4

Tuesday — 5

Wednesday — 6

Thursday — 7

Friday — 8
- USDA reports: WASDE, Crop Production

Saturday — 9

My 2013 Beans: _____ acres X _____ bu/acre X $ _____ per bushel + $ _____
My 2013 Corn: _____ acres X _____ bu/acre X $ _____ per bushel + $ _____

In the USDA Crop Production Report this Month:
Papayas, Citrus Fruits, Sugarcane

		Crop Year		
Item	**2010**	**2011**	**2012** Estimated (as of Sept 2012)	**2013** Projected (as of Mar 8, 2013)

US Corn Supply and Use
(millions of bushels)

Item	2010	2011	2012 Estimated (as of Sept 2012)	2013 Projected (as of Mar 8, 2013)
Area planted (mil. ac.)	*88.2*	*91.9*	*96.4*	
Area harvested (mil. ac.)	*81.4*	*84.0*	*87.4*	
Yield per acre (bu)	*152.8*	*147.2*	*122.8*	
Beginning stocks	1,708	1,128	1,181	
Production	12,447	12,358	10,727	
Imports	28	25	75	
Total Supply	**14,182**	**13,511**	**11,983**	
Feed and residual	4,793	4,400	4,150	
Food, seed & industrial*	6,428	6,390	5,850	
Exports	1,834	1,540	1,250	
Total Use	**13,055**	**12,330**	**11,250**	
ENDING STOCKS	**1,128**	**1,181**	733	

* This includes:

Corn used for ethanol	*5,021*	*5,000*	*4,500*	

 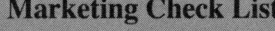 

Marketing Check List

❑ Note the average price for December Corn and November Soybeans at the end of February. This is an important price level to be aware of when making all future new crop pricing decisions.

US Soybean Supply and Use
(in millions of bushels)

Item	Crop Year			
	2010	2011	2012 Estimated (as of Sept 2012)	2013 Projected (as of Mar 8, 2013)
Area planted (mil. ac.)	77.4	75.0	76.1	
Area harvested (mil. ac.)	76.6	73.6	74.6	
Yield per acre (bu)	43.5	41.5	35.3	
Beginning stocks	151	215	130	
Production	3,329	3,056	2,634	
Imports	14	16	20	
Total Supply	**3,495**	**3,287**	**2,785**	
Crushings	1,648	1,705	1,500	
Exports	1,501	1,360	1,055	
Seed	87	88	89	
Residual	44	3	25	
Total Use	**3,280**	**3,157**	**2,670**	
ENDING STOCKS	**215**	**130**	**115**	

 Dutch Tip

"The Right Stuff": Buy good land. You only pay for it once. Buy poor land and you pay for it forever.

10-16

5-Year Seasonal Odds this Week		
	change	reliability
Corn	+9.4¢ ⇧	80%
Soybeans	+10.0¢ ⇧	80%

New Crop This Week

	Hi	Lo	Close	12 Close
December Corn				$5.74
November Soybeans				$13.28

Old Crop This Week

	Hi	Lo	Close	12 Close
May Corn				$6.73
May Soybeans				$13.74

Sunday 10
- Daylight Savings Time begins

Monday 11

Tuesday 12
- Al's "Second Tuesday" Webinar

Wednesday 13

Thursday 14
- Last Trading Day: Mar Soybeans, Mar Corn

Friday 15

Saturday 16

My 2013 Beans: _____ acres X _____ bu/acre X $ _____ per bushel + $ _____
My 2013 Corn: _____ acres X _____ bu/acre X $ _____ per bushel + $ _____

Three Suggestions for Marketing These Days

What should farmers do differently in this new marketing era? I have three suggestions.

 Stay informed. I believe most farmers make the best decisions early in the day. Check the markets, read some early morning advice, and then head out to the field or shop.

 Use all of the marketing tools that are available. Thirty years ago you had two choices: sell cash or forward contract. Now you can use hedges, hedge-to-arrive contracts, and options. By using all of the tools, you greatly increase the odds you will make the right decision.

 Call your offers in. Having a resting offer in place above the market where you will make the sale is likely to get your crop sold at night, on the open of trade, or at a time when you are out in the field. Work with a grain broker you can trust and place your orders.

Marketing Check List

 ❑ Finalize your crop plans for 2013 and get all crop insurance contracts signed.

Acres of Corn PLANTED as of June 30, 2012 (1,000 acres)			
State	2010	2011	2012 (June 30, 2012)
AL	270	270	290
AZ	45	55	65
AR	390	560	660
CA	610	630	610
CO	1,330	1,500	1,420
CT	26	27	28
DE	180	190	195
FL	60	65	70
GA	295	345	335
ID	320	350	380
IL	12,600	12,600	13,000
IN	5,900	5,900	6,200
IA	13,400	14,100	14,000
KS	4,850	4,900	4,700
KY	1,340	1,380	1,600
LA	510	580	570
ME	28	29	31
MD	500	500	490
MA	17	17	17
MI	2,400	2,500	2,600
MN	7,700	8,100	8,700
MS	750	810	840
MO	3,150	3,300	3,600
MT	80	77	100
NE	9,150	9,850	9,900
NV	4	8	8
NH	15	15	14
NJ	80	90	90
NM	140	125	125
NY	1,050	1,100	1,160
NC	910	870	850
ND	2,050	2,230	3,400
OH	3,450	3,400	3,900
OK	370	380	370
OR	70	83	85
PA	1,350	1,420	1,460
RI	2	2	1
SC	350	360	320
SD	4,550	5,200	6,000
TN	710	790	930
TX	2,300	2,050	1,900
UT	70	85	85
VT	92	90	94
VA	490	490	510
WA	200	195	200
WV	48	48	52
WI	3,900	4,150	4,350
WY	90	105	100

17-23

5-Year Seasonal Odds this Week		
	change	reliability
Corn	+1.0¢ ⇧	40%
Soybeans	+0.6¢ ⇧	20%

New Crop This Week

	Hi	Lo	Close	12 Close
December Corn				$5.58
November Soybeans				$13.22

Old Crop This Week

	Hi	Lo	Close	12 Close
May Corn				$6.47
May Soybeans				$13.66

Sunday — 17
- St. Patricks Day

Monday — 18
- Last Delivery Day: Mar Soybeans, Mar Corn

Tuesday — 19

Wednesday — 20
- First Day Of Spring

Thursday — 21

Friday — 22
- USDA report: Cattle On Feed

Saturday — 23

My 2013 Beans: _____ acres X _____ bu/acre X $ _____ per bushel + $ _____
My 2013 Corn: _____ acres X _____ bu/acre X $ _____ per bushel + $ _____

12 BIG WORDS O' WEATHER
From DT, Al's Favorite Weather Guy

#3. TELECONNECTION

The atmosphere is not uniform. Even though it consists of air that we can see, air acts like a fluid. This means that when there are big temperature differences, air will shift and move. However, it never has a chance to mix completely or evenly. As a result, there are often irregular areas of warm or cold air over certain portions of the atmosphere where you might not expect them.

For example, on some weather maps a meteorologist can see a buildup of warm air over the West Coast of North America. This buildup is often referred to as a "ridge" within the jet stream, and it can lead to a change in the pattern over central and eastern Canada and the U.S.

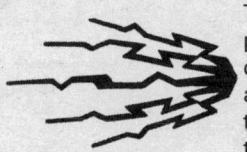

There are many different types of anomalous weather patterns that develop over the northern hemisphere during all four seasons. Sometimes the buildup of an anomalous weather pattern in one portion of the globe means that something is going to occur thousands of miles away. This relationship is called a 'teleconnection', from the word root 'tele-', which means far. Many types of 'far connections', or teleconnections, have been identified and catalogued by meteorologists and climatologists over the past 20 years.

24-30

5-Year Seasonal Odds this Week		
	change	reliability
Corn	+17.0¢ ⇧	60%
Soybeans	+30.0¢ ⇧	80%

New Crop This Week

	Hi	Lo	Close	12 Close
December Corn				$5.40
November Soybeans				$13.58

Sunday 24

Monday 25

Tuesday 26

Wednesday 27

Old Crop This Week

	Hi	Lo	Close	12 Close
May Corn				$6.44
May Soybeans				$14.03

Thursday 28
- USDA report: Quarterly Hogs and Pigs, Grain Stocks, Prospective Plantings
- Al's "Last Thursday" Webinar

Friday 29
- CBOT CLOSED: Good Friday

	Close Today	Close Last Month	Difference
Corn			
Soybeans			

Saturday 30

My 2013 Beans: _____ acres X _____ bu/acre X $ _____ per bushel + $ _____
My 2013 Corn: _____ acres X _____ bu/acre X $ _____ per bushel + $ _____

Al-ism

"USDA reports produce more drama than Hollywood, and they're all rated G."

How much will we plant? Acres *intended* to be planted, as of March 28 (1,000 acres)					
Crop	2011	2012	% of prior year	March 28, 2013	% of prior year
Corn	92,178	95,864	104		
Soybeans	76,609	73,902	99		
Hay, All (harvested)	58,973	57,348	97		
Wheat, All	58,021	55,908	96		
Cotton, All	12,565	13,155	105		

The end of March "Prospective Plantings" report from the USDA is one of the biggest market movers of the year. In 2013, when farmers are deciding what to plant, they will carefully evaluate new crop futures prices and input costs before turning a wheel.

Marketing Check List

❑ Get at least 50% of the basis locked in on all May hedges.

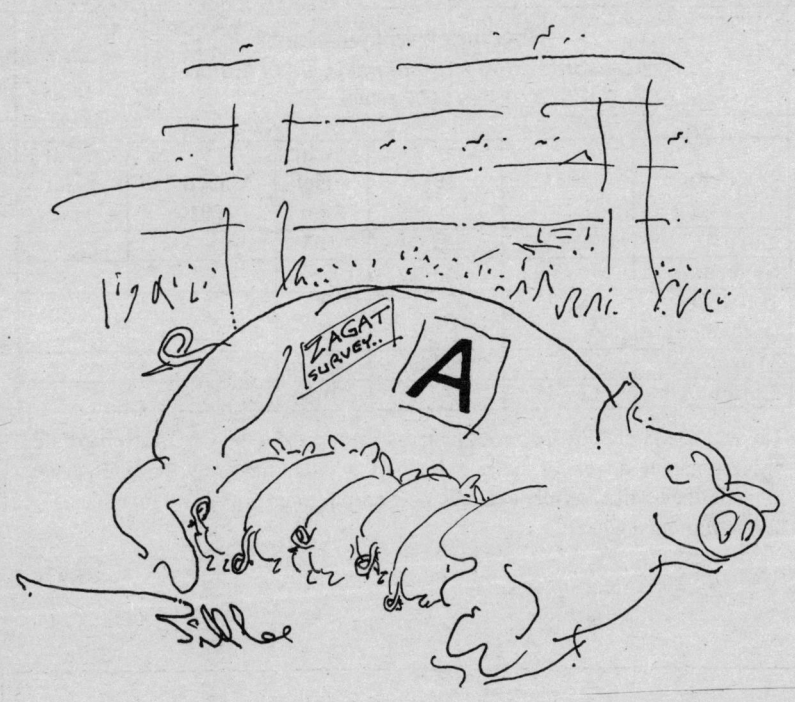

What's left?

*US **grain stocks** stored on and off the farm (in thousands of bushels)*

	Date	2010	2011	2012 (as of Sept 2012)	2013 (as of March 29, 2013)
Corn	Mar 1	7,693,787	6,523,228	6,023,356	
	Jun 1	4,310,071	3,669,838	3,148,540	
	Sep 1	1,707,787	1,127,645	988,403	
	Dec 1	10,056,769	9,647,466		
Soybeans	Mar 1	1,270,068	1,248,800	1,374,488	
	Jun 1	571,123	619,283	667,475	
	Sep 1	150,885	215,013	169,417	
	Dec 1	2,278,084	2,369,885		

Inventory of all US Hogs & Pigs
(1,000 head)

Date of tally	2010	2011	2012	2013	2013 as % of	
					2011	2012
Mar 1	63,568	63,684	64,937			
June 1	64,650	65,320	65,759			
Sept 1	65,971	67,234	67,472			
Dec 1	64,925	66,361				

31-6

5-Year Seasonal Odds this Week

	change	reliability
Corn	+3.9¢ ⇧	60%
Soybeans	+21.0¢ ⇧	80%

New Crop This Week

	Hi	Lo	Close	12 Close
December Corn				$5.50
November Soybeans				$13.81

Old Crop This Week

	Hi	Lo	Close	12 Close
May Corn				$6.58
May Soybeans				$14.34

Sunday — 31
- Easter

Monday — 1
- USDA report: Crop Progress

Tuesday — 2

Wednesday — 3

Thursday — 4

Friday — 5

Saturday — 6

My 2013 Beans: _____ acres X _____ bu/acre X $ _____ per bushel + $ _____
My 2013 Corn: _____ acres X _____ bu/acre X $ _____ per bushel + $ _____

Al's Fearless Forecast

Al says: "Oat futures will trade over $5.00
per bushel by July of 2013."

Dust Explosions in Threshing Machines (Bureau of Chemistry and Soils). 1 reel - 804 feet.

Explosions and fires in threshing machines, their causes and results; use of preventive devices—suction fans, fire extinguishers, wire systems; experiments with dust from starch, flour, sugar, coal, and sulphur. Of general interest to wheat growers.

Construction of a Concrete Silo (Bureau of Dairy Industry). 1 reel - 985 feet.

The practicability of building a concrete silo with labor available on the farm, and the actual building, showing the various steps. Of general interest for rural communities.

USDA, *Division of Motion Pictures, 1934*

Marketing Check List

☐ Write down the USDA March acreage projections for 2013 and compare to 2012.

7-13

5-Year Seasonal Odds this Week		
	change	reliability
Corn	-7.1¢ ⬇	60%
Soybeans	+9.8¢ ⬆	80%

New Crop This Week

	Hi	Lo	Close	12 Close
December Corn				$5.37
November Soybeans				$13.61

Old Crop This Week

	Hi	Lo	Close	12 Close
May Corn				$6.29
May Soybeans				$14.37

Sunday · 7

Monday · 8
- USDA report: Crop Progress

Tuesday · 9
- Al's "Second Tuesday" Webinar

Wednesday · 10
- USDA reports: WASDE, Crop Production

Thursday · 11

Friday · 12

Saturday · 13

My 2013 Beans: _____ acres X _____ bu/acre X $ _____ per bushel + $ _____
My 2013 Corn: _____ acres X _____ bu/acre X $ _____ per bushel + $ _____

Who GROWS the most (and least) corn?
(in millions of metric tons, as of September 2012)

Country	2010-2011	2011-2012	2012-2013
US	316.17	313.92	272.49
China	177.25	192.78	200.00
Europe (EU-27)	55.93	65.40	51.14
Brazil	57.40	72.73	70.00
SE Asia	23.01	25.38	25.60
Mexico	21.06	18.10	21.50
FSU-12	18.49	33.69	32.06
Argentina	25.20	21.00	28.00
South Africa	10.92	11.50	13.50
Canada	11.71	10.69	11.70
Ukraine	11.92	22.84	21.00
Egypt	6.50	5.50	5.80
So. Korea	0.07	0.07	0.08
Japan	0.00	0.00	0.00
WORLD	**830.77**	**876.68**	**841.06**

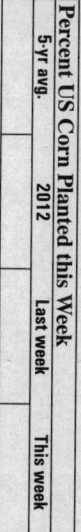

USDA CROP PROGRESS REPORT

Percent US Corn Planted this Week

5-yr avg.	2012	Last week	This week

Marketing Check List
☐ Check the quality of your stored grain. As temperatures move up, problems can develop.

12 Big Words o' Weather
From DT, Al's Favorite Weather Guy

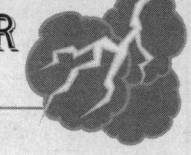

#4. PDO

A very important teleconnection (see Word o' Weather #3) is the PDO, or "Pacific Decadal Oscillation". A PDO has a positive or negative phase. The weather patterns resulting from a positive PDO is vastly different from a weather pattern based on a negative PDO.

The PDO phase--whether positive or negative--is determined by sea surface temperatures. If the sea surface temps are colder than normal, this is a negative PDO. A negative PDO has a typical pattern that will be more likely to occur for one or several months. For example, starting during the winter of 2011-12, the PDO was extremely negative. This continued up through summer 2012. As a result, there was a persistent trough in the jet stream over the West Coast and a trough over the southeast coast. This in turn prevented any significant cold air outbreaks from occurring over the central and eastern U.S. during the winter of 2011-12. It also ensured that most of the Midwest and the Plains states would stay hot and dry during the summer months, while the Pacific Northwest and south-central Canada saw above-normal rainfall.

When a teleconnection like the PDO reaches an extreme or record-setting level--and holds that extreme for several weeks or months--it usually means the current season is likely to experience exceptional conditions as well. During the winter, spring, and summer months of 2012, the PDO was extremely negative and at times reached a record low reading. Not surprisingly, the drought and the heat over the Plains and the Midwest was extreme as well.

14-20

5-Year Seasonal Odds this Week		
	change	reliability
Corn	-11.0¢ ⇩	80%
Soybeans	+5.4¢ ⇧	60%

New Crop This Week

	Hi	Lo	Close	12 Close
December Corn				$5.37
November Soybeans				$13.56

Old Crop This Week

	Hi	Lo	Close	12 Close
May Corn				$6.13
May Soybeans				$14.47

Sunday — 14

Monday — 15
- USDA report: Crop Progress

Tuesday — 16

Wednesday — 17

Thursday — 18

Friday — 19
- USDA report: Cattle On Feed

Saturday — 20

My 2013 Beans: _____ acres X _____ bu/acre X $ _____ per bushel + $ _____
My 2013 Corn: _____ acres X _____ bu/acre X $ _____ per bushel + $ _____

Percent US Corn Planted this Week			
5-yr avg.	2012	Last week	This week

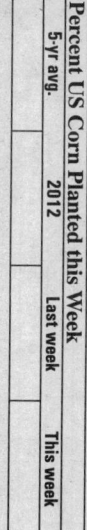

Al's Fearless Forecast

Al says: "Greece will be out of the Euro
by the third quarter of 2013."

Al-ism

"Who's that diverse financial advisory committee for that global
manufacturer? Oh... that's your marketing team!"

Marketing Check List

❏ Get all cash basis set for May hedges.

21-27

5-Year Seasonal Odds this Week		
	change	reliability
Corn	+25.0¢ ⇧	100%
Soybeans	+16.0¢ ⇧	60%

Sunday   21

Monday   22
- USDA report: Crop Progress

Tuesday   23

Wednesday   24

Thursday   25
- Al's "Last Thursday" webinar

Friday   26

Saturday   27

New Crop This Week

	Hi	Lo	Close	12 Close
December Corn				$5.39
November Soybeans				$13.62

Old Crop This Week

	Hi	Lo	Close	12 Close
May Corn				$6.53
May Soybeans				$14.97

My 2013 Beans: _____ acres X _____ bu/acre X $ _____ per bushel + $ _____
My 2013 Corn: _____ acres X _____ bu/acre X $ _____ per bushel + $ _____

Percent US Corn Planted this Week			
5-yr avg.	2012	Last week	This week

Marketing Check List

☐ Use a mid-month rally to make additional cash grain sales.

28-4

New Crop This Week

	Hi	Lo	Close	12 Close
December Corn				$5.24
November Soybeans				$13.67

Old Crop This Week

	Hi	Lo	Close	12 Close
May Corn				$6.62
May Soybeans				$14.75

Sunday — 28

Monday — 29

- USDA report: Crop Progress

Tuesday — 30

	Close Today	Close Last Month	Difference
Corn			
Soybeans			

Wednesday — 1

Thursday — 2

Friday — 3

Saturday — 4

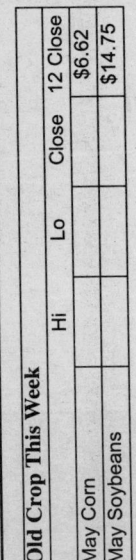

My 2013 Beans: _____ acres X _____ bu/acre X $ _____ per bushel + $ _____
My 2013 Corn: _____ acres X _____ bu/acre X $ _____ per bushel + $ _____

In the USDA Crop Production Report this Month:

Hay, Wheat, Cotton, Tobacco, Almonds, Bananas, Guavas, Papayas, Peaches, Taro, Citrus Fruits, Potatoes

HYACINTHS.

USDA CROP PROGRESS REPORT			
Percent US Corn Planted this Week			
5-yr avg.	2012	Last week	This week

 Dutch Tip

"Better by the Bushel": Don't complain about input costs. Think of your fertilizer and fuel costs in the number of bushels it takes you to pay for them.

Marketing Check List

❑ Complete all cash wheat sales.

Who GROWS the most (and least) wheat?
(in millions of metric tons, as of Sept 2012)

Country	2010-11	2011-12	2012-13
EU-27	135.9	137.4	132.4
China	115.2	117.9	118.0
FSU-12	81.1	114.4	79.0
India	80.8	86.9	93.9
Russia	41.5	56.2	39.0
United States	**60.1**	**54.4**	**61.7**
Canada	23.2	25.3	27.0
Pakistan	23.9	24.2	23.0
Australia	27.9	29.5	26.0
Ukraine	16.8	22.1	15.5
North Africa	15.9	18.4	17.0
Selected Mideast	19.6	17.6	17.6
Kazakhstan	9.6	22.7	10.5
Argentina	16.7	15.0	11.5
Brazil	5.9	5.8	5.0
SE Asia	0.0	0.0	0.0
WORLD	**651.9**	**695.0**	**658.73**

Who GROWS the most (and least) soybeans?
(in millions of metric tons, as of Sept 2012)

Country	2010	2011	2012
US	90.6	83.2	71.7
Brazil	75.5	66.5	81.0
Argentina	49.0	41.0	55.0
China	15.1	13.5	12.6
Europe (EU-27)	1.1	1.3	1.1
Japan	0.2	0.2	0.2
Mexico	0.2	0.2	0.2
WORLD	**264.7**	**237.1**	**258.13**

12 Big Words o' Weather
From DT, Al's Favorite Weather Guy

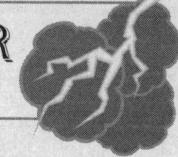

#5. Arctic Oscillation

Another very important teleconnection (see Words o' Weather #3) is the AO, or Arctic Oscillation. However, unlike the PDO (see Words o' Weather #4), the AO is not defined by ocean surface temperatures. Instead, the AO is determined by the amount of warm or cold air "pockets" over the Arctic regions of the northern hemisphere.

As with the PDO, the AO can be positive or negative. When the Arctic oscillation is negative, it strongly supports an active, stormy, cold spring with late season snows and stronger-than-normal cold fronts, which bring stronger and more frequent rains to the Plains and the Midwest. It also strongly supports the active and stormy winter pattern across all of the country, especially east of the Rockies and all way to the east coast.

On the other hand, when the Arctic oscillation is positive, it means most of the extreme cold air in Canada is trapped in central and Northern Canada and is having a great deal of difficulty moving out of Canada into the continental US.

The AO was extremely positive during the winter of 2011-2012. In fact, it was one of the top five most-positive Arctic oscillation readings seen since 1950. As I stated above, extreme teleconnections (such as an AO) often result in extreme weather patterns. Thus, when we had an extremely positive Arctic oscillation last winter, spring, and early summer, we saw a prolonged period of above-normal temperatures and below-normal rainfall.

May

5-11

New Crop This Week

	Hi	Lo	Close	12 Close
December Corn				$5.05
November Soybeans				$13.21

Old Crop This Week

	Hi	Lo	Close	12 Close
May Corn				$6.08
May Soybeans				$14.04

Sunday   5

Monday   6
- USDA report: Crop Progress

Tuesday   7

Wednesday   8

Thursday   9

Friday   10
- USDA reports: WASDE, Crop Production

Saturday   11

My 2013 Beans: _____ acres X _____ bu/acre X $ _____ per bushel + $ _____
My 2013 Corn: _____ acres X _____ bu/acre X $ _____ per bushel + $ _____

How did **stock markets** do in **2012**... around the world?		
Country	**Stock Index**	**% Gain or loss ***
Egypt	Case 30	+ 52.1
Turkey	ISE	+ 47.8
Pakistan	KSE	+ 34.0
Thailand	SET	+ 30.4
Poland	WIG	+ 21.7
Germany	DAX	+ 20.5
Hong Kong	Hang Seng	+ 20.1
India	BSE	+ 20.1
Denmark	OMXCB	+ 19.6
Greece	Athex Comp	+ 19.2
Mexico	IPC	+ 18.2
Norway	OSEAX	+ 13.6
South Korea	KOSPI	+ 12.6
Australia	All Ord.	+ 12.4
Chile	IGPA	+ 12.4
Malaysia	KLSE	+ 11.8
Belgium	Bel 20	+ 11.5
Sweden	OMXS30	+ 9.4
Indonesia	JSX	+ 7.5
Czech Republic	PX	+ 7.1
Taiwan	TWI	+ 7.0
Britain	FTSE 100	+ 6.9
Israel	TA-100	+ 6.3
France	CAC 40	+ 6.0
US	**DJIA**	**+ 5.9**
Euro area	DJ STOXX 50	+ 5.1
Netherlands	AEX	+ 4.7
Canada	S&P TSX	+ 4.6
Russia	RTS	+ 3.3
Japan	Nikkei 225	+ 2.2
Italy	FTSE/MIB	- 0.4
China	SSEA	- 3.5
Brazil	BVSP	- 5.4
Spain	Madrid SE	- 11.7
Argentina	MERV	- 13.2

* Gain (in US dollar terms) from Jan 1, 2012 to Nov 7, 2012

As much as we want to be the winner on any list, economic growth around the globe is very good for US farmers. Countries with growing economies have citizens who want better food and more meat--and they have the cash to pay for it.

USDA CROP PROGRESS REPORT

Percent US Corn Planted this Week

5-yr avg.	2012	Last week	This week

Marketing Check List

☐ Get the last of your cash basis set on all July corn and soybean hedges.

12-18

5-Year Seasonal Odds this Week		
	change	reliability
Corn	+28.0¢ ⇧	100%
Soybeans	+12.0¢ ⇧	60%

New Crop This Week

		12 Close	Close	Lo	Hi
December Corn		$5.37			
November Soybeans		$12.88			

Old Crop This Week

		12 Close	Close	Lo	Hi
July Corn		$6.36			
July Soybeans		$14.05			

Sunday — 12
- Mothers Day

Monday — 13
- USDA report: Crop Progress

Tuesday — 14
- Last Trading Day: May Soybeans, May Corn
- Al's "Second Tuesday" Webinar

Wednesday — 15

Thursday — 16
- Last Delivery Day: May Soybeans, May Corn

Friday — 17
- USDA report: Cattle On Feed

Saturday — 18

My 2013 Beans: _____ acres X _____ bu/acre X $ _____ per bushel + $ _____
My 2013 Corn: _____ acres X _____ bu/acre X $ _____ per bushel + $ _____

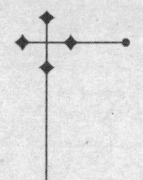

♭reakfast

Broccoli pancakes,

Brussels sprout pie,

Lima bean donuts,

Onions on rye.

Daffodil pudding,

Liverwurst bread,

Dad's cooking breakfast—

I'M STAYING IN BED!

USDA CROP PROGRESS REPORT			
Percent US Corn Planted this Week			
5-yr avg.	2012	Last week	This week

Mother Takes a Vacation (Extension Service). 2 reels - 1,206 feet.

Depicts the activities of farm women's vacation camps conducted by State extension services in the West. Photographed in Montana and Idaho.

USDA, *Division of Motion Pictures, 1934*

 ♭l's ♭earless ♭orecast

Al says: "New all-time highs in the hog market
by the third quarter of 2013."

Marketing Check List

❑ Use any mid-month rally to increase cash corn and
soybean sales up to 80% or more.

12 BIG WORDS O' WEATHER
From DT, Al's Favorite Weather Guy

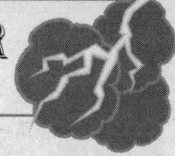

#6. NAO

Another very important teleconnection (see Words o' Weather #3) is the NAO, or North American Oscillation. The main area of the NAO is over Northeast Canada, Greenland and Iceland. When there is a bubble of warm air in the upper levels of the atmosphere--at the Jet stream level--over one or more of these areas, the NAO is considered to be in the negative, or warm, phase.

As a general rule, a negative NAO results in a colder and wetter-than-normal pattern over much of the Midwest and the East Coast of the continental US. A negative NAO also strongly impacts weather in western and Central Europe. However there is also a cold or positive phase of the NAO. This occurs when there is a large pool of colder-than-normal air at the Jet stream level over northeastern Canada, Greenland, and/or Iceland.

A positive NAO has a strong correlation to warmer-than-normal patterns over the central and eastern portions of the continental US. and over western and Central Europe.

Generally the NAO pattern is strongest when it is aligned with the AO (Arctic oscillation pattern, see Words o' Weather #5). A combination of a negative NAO and a negative AO almost always has a profound and significant impact on weather patterns over North America and in Europe .

There are times when the NAO and AO are not in the same phase. When that happens, the influences of these two weather patterns are generally diminished.

During the summer months a negative NAO will allow a cold front to drop further into the Midwest and bring more significant rain, especially when the cold fronts stall across the Midwest and/or the central Plains. During the summer and severe drought of 2012, both the AO and NAO were strongly positive. This helped keep cold fronts north of the Great Lakes and  the U.S./Canada border. In turn, this restricted the rain chances during May, June, and July. It changed in August 2012. Even though most forecasters called for continuation of the drought, the AO turned neutral and the NAO turned negative. This allowed the Jet stream to drop further south, driving cold fronts into the Midwest, breaking the heat, and bringing significantly more rain than the Midwest had seen over the past five months.

19-25

New Crop This Week

	Hi	Lo	Close	12 Close
December Corn				$5.22
November Soybeans				$12.89

Old Crop This Week

	Hi	Lo	Close	12 Close
July Corn				$5.79
July Soybeans				$13.82

Sunday 19

Monday 20
- USDA report: Crop Progress

Tuesday 21

Wednesday 22

Thursday 23

Friday 24

Saturday 25

My 2013 Beans: _____ acres X _____ bu/acre X $ _____ per bushel + $ _____
My 2013 Corn: _____ acres X _____ bu/acre X $ _____ per bushel + $ _____

Percent US Corn Planted this Week				Percent US Corn Good to Excellent this Week			
5-yr avg.	2012	Last week	This week	5-yr avg.	2012	Last week	This week
Good							
Excellent							

Al-ism

"Sell when markets are going up, or you'll be crying--and sell-
ing--as they go down."

Al's Fearless Forecast

Al says: "The S&P 500 will drop below 1300
by the third quarter of 2013."

Marketing Check List

❑ Get hedges in place on at least 10-20% of your new
crop corn and soybeans.

26-1

5-Year Seasonal Odds this Week		
	change	reliability
Corn	+1.8¢ ⇧	40%
Soybeans	+26.0¢ ⇧	60%

New Crop This Week

	Hi	Lo	Close	12 Close
December Corn				$5.10
November Soybeans				$12.58

Old Crop This Week

	Hi	Lo	Close	12 Close
July Corn				$5.52
July Soybeans				$13.44

Sunday 26

Monday 27
• CBOT Closed: Memorial Day

Tuesday 28
• USDA report: Crop Progress

Wednesday 29

Thursday 30
• Al's "Last Thursday" Webinar

Friday 31

	Close Today	Close Last Month	Difference
Corn			
Soybeans			

Saturday 1

My 2013 Beans: _____ acres X _____ bu/acre X $ _____ per bushel + $ _____
My 2013 Corn: _____ acres X _____ bu/acre X $ _____ per bushel + $ _____

What's in the **Dow Jones**?

3M	IBM
Alcoa	Johnson & Johnson
American Express	JPMorgan Chase
AT&T	McDonalds
Bank of America	Merck
Boeing	Microsoft
Caterpillar	Pfizer
Chevron	Procter & Gamble
Cisco Systems	Coca-Cola
duPont	Travelers Companies
Exxon Mobil	United Technologies
General Electric	UnitedHealth Group
Hewlett-Packard	Verizon
Home Depot	Wal-Mart
Intel	Walt Disney

Percent US Corn Good to Excellent this Week	5-yr avg.	2012	Last week	This week
Good				
Excellent				

You've heard about "the Dow Jones" (the Dow Jones Industrial Average, or DJIA)... but what is it? The DJIA is a set of 30 major companies that have been chosen to, as a group, represent the health of the US economy. Together, they are called an 'index'.

 Dutch Tip

"Fill 'Er Up": Buy your fall fuel needs ahead in June. It just about always pays off. It sure did in 2012.

Marketing Check List

☐ Get puts bought on at least 20-30% of your 2013 crop corn and soybeans.

2-8

5-Year Seasonal Odds this Week		
	change	reliability
Corn	+30.0¢ ⇧	80%
Soybeans	+38.0¢ ⇧	80%

New Crop This Week

	Hi	Lo	Close	12 Close
December Corn				$5.44
November Soybeans				$13.32

Old Crop This Week

	Hi	Lo	Close	12 Close
July Corn				$5.98
July Soybeans				$14.26

Sunday 2

Monday 3
- USDA report: Crop Progress

Tuesday 4

Wednesday 5

Thursday 6

Friday 7

Saturday 8

My 2013 Beans: _____ acres X _____ bu/acre X $ _____ per bushel + $ _____
My 2013 Corn: _____ acres X _____ bu/acre X $ _____ per bushel + $ _____

In the USDA Crop Production
Report this Month:
Barley, Oats, Wheat, Tobacco, Lentils, Peas, Almonds, Apricots, Grapes, Papayas, Peaches, Citrus Fruits, Potatoes

USDA CROP PROGRESS REPORT

Percent US Corn Good to Excellent this Week	5-yr avg.	2012	Last week	This week
Good				
Excellent				

Percent US Soybeans Good to Excellent this Week	5-yr avg.	2012	Last week	This week
Good				
Excellent				

Al's Fearless Forecast

Al says: "Crude oil will drop below $80 per barrel by August of 2013."

Marketing Check List

❏ Check on fall fuel costs and get some fall fuel and LP bought.

June

9-15

New Crop This Week

	Hi	Lo	Close	12 Close
December Corn				$5.06
November Soybeans				$13.14

Old Crop This Week

	Hi	Lo	Close	12 Close
July Corn				$5.80
July Soybeans				$13.76

Sunday 9

Monday 10
- USDA report: Crop Progress

Tuesday 11
- Al's "Second Tuesday" Webinar

Wednesday 12
- USDA reports: WASDE, Crop Production

Thursday 13

Friday 14

Saturday 15
- LOW: Dec 2014 Corn, $4.99 on June 15, 2012
- LOW: Dec 2015 Corn, $5.03 on June 15, 2012

My 2013 Beans: _____ acres X _____ bu/acre X $ _____ per bushel + $ _____
My 2013 Corn: _____ acres X _____ bu/acre X $ _____ per bushel + $ _____

	5-yr avg.	2012	Last week	This week
Percent US Corn Good to Excellent this Week				
Excellent				
Good				
Percent US Soybeans Good to Excellent this Week				
Excellent				
Good				

My Garden

My garden is growing electric guitars
with lavender trumpets and musical stars.
A purple piano is sprouting her keys,
and saxophone swans are performing in trees.
The tambourine tulips are humming a tune.
A silkworm is drumming inside her cocoon.
The violin violets are handing out treats.
My garden is filled with harmonious beets.

Marketing Check List

❑ Use a mid-month rally to complete cash corn and soybean sales.

12 BIG WORDS O' WEATHER
From DT, The Weather Guy

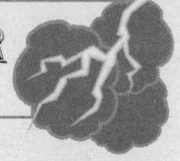

#7. PNA

Another very important teleconnection (see Words o' Weather #3) is the PNA, or Pacific North American.

The PNA pattern concerns the area along the eastern Pacific and Western North America, including the west coast of the US and Canada. When there is a bubble of warm air in the jet stream over the west coast, the PNA is in the positive phase. A positive PNA deflects the jet stream. Normally, the jet stream circles the globe in the middle latitudes. But in a positive PNA, the jet stream is diverted, and drops down from Northwest Canada and into the Plains, the Midwest, or the east coast. During the winter, this can bring in arctic air masses, which sometimes help set the stage for snowstorms. During the summer and spring, a positive PNA pattern will bring in colder-than-normal air masses. That makes cold fronts drive deeper into the Plains and Midwest, and raises the chances for rain.

On the other hand, a pocket of colder-than-normal air over western North America means a negative PNA. This is also very important, but for different sections of the country. A negative PNA will mean above-normal rainfall and generally cooler-than-normal temperatures on the West Coast and the Rockies. A negative PNA pattern will often (but not always) result in low pressure areas from the West Coast up towards the upper Plains, Manitoba, or the western Great Lakes. However, over the lower Plains and the eastern third of the US, a negative PNA pattern can result in above-normal temperatures and drier-than-normal conditions.

10 Things Good Marketers Do

They...

1. write down their marketing plan and review it on a regular basis.

2. make marketing a year-round project.

3. stay updated on market news, price movement, and basis change
   each day.

4. review the markets early each morning.

5. make decisions early in the day, before the market opens.

6. think through their marketing plan.

7. don't panic when they hear negative news during the day.

8. have a positive attitude about farming.

9. have a positive attitude about marketing.

10. feel that marketing is an exciting challenge, like a game they can learn and tackle, season after season.

June

16-22

New Crop This Week

	12 Close	Close	Lo	Hi
December Corn	$5.54			
November Soybeans	$13.75			

Old Crop This Week

	12 Close	Close	Lo	Hi
July Corn	$5.91			
July Soybeans	$14.43			

Sunday — 16
- Father's Day
- HIGH: Dec 2010 Corn, $7.03 on June 16, 2008

Monday — 17
- USDA report: Crop Progress

Tuesday — 18

Wednesday — 19

Thursday — 20

Friday — 21
- First Day Of Summer
- USDA report: Cattle On Feed

Saturday — 22

My 2013 Beans: _____ acres X _____ bu/acre X $ _____ per bushel + $ _____
My 2013 Corn: _____ acres X _____ bu/acre X $ _____ per bushel + $ _____

Percent US Corn Good to Excellent this Week	5-yr avg.	2012	Last week	This week
Good				
Excellent				
Percent US Soybeans Good to Excellent this Week	5-yr avg.	2012	Last week	This week
Good				
Excellent				

Marketing Check List

☐ Increase new crop hedges for corn and soybeans up to 30-40%.

June

23-29

5-Year Seasonal Odds this Week		
	change	reliability
Corn	+5.7¢ ⇧	40%
Soybeans	+39.0¢ ⇧	100%

New Crop This Week

	Hi	Lo	Close	12 Close
December Corn				$6.35
November Soybeans				$14.27

Old Crop This Week

	Hi	Lo	Close	12 Close
July Corn				$6.72
July Soybeans				$15.13

Sunday — 23

Monday — 24
- USDA report: Crop Progress

Tuesday — 25

Wednesday — 26

Thursday — 27
- Al's "Last Thursday" Webinar

Friday — 28
- USDA report: Quarterly Hogs and Pigs, Acreage, Grain Stocks

	Close Today	Close Last Month	Difference
Corn			
Soybeans			

Saturday — 29
- LOW: Dec 2010 Corn, $3.43 on June 29, 2010
- LOW: Dec 2013 Corn, $3.98 on June 29, 2010

My 2013 Beans: _____ acres X _____ bu/acre X $ _____ per bushel + $ _____
My 2013 Corn: _____ acres X _____ bu/acre X $ _____ per bushel + $ _____

USDA CROP PROGRESS REPORT

Percent US Corn Good to Excellent this Week

	5-yr avg.	2012	Last week	This week
Good				
Excellent				

Percent US Soybeans Good to Excellent this Week

	5-yr avg.	2012	Last week	This week
Good				
Excellent				

How much did we plant?
Acres **planted** as of June 28
(1,000 acres)

Crop	2011	2012	% of prior year	2013	% of prior year
Corn	91,921	96,405	105		
Soybeans	74,976	76,080	102		
Hay, All (harvested)	55,633	57,669	104		
Wheat, All	54,409	56,017	103		
Cotton, All	14,735.4	12,635.0	86		

Marketing Check List

❑ Buy new crop puts on an additional 10-20% of your 2012 corn and soybean crops.

30-6

5-Year Seasonal Odds this Week		
	change	reliability
Corn	+9.4¢ ⇧	60%
Soybeans	+11.0¢ ⇧	60%

New Crop This Week		12 Close	Close	Lo	Hi
	December Corn	$6.93			
	November Soybeans	$15.06			

Old Crop This Week		12 Close	Close	Lo	Hi
	July Corn	$7.43			
	July Soybeans	$16.20			

Sunday — 30

Monday — 1
- USDA report: Crop Progress

Tuesday — 2

Wednesday — 3
- HIGH: Nov 2010 Soybeans, $15.55 on July 3, 2008
- HIGH: Nov 2011 Soybeans, $15.50 on July 3, 2008

Thursday — 4
- CBOT Closed: Independence Day

Friday — 5

Saturday — 6

My 2013 Beans: _____ acres X _____ bu/acre X $ _____ per bushel + $_____
My 2013 Corn: _____ acres X _____ bu/acre X $ _____ per bushel + $_____

In the USDA Crop Production Report this Month:
Barley, Oats, Wheat, Tobacco, Lentils, Peas, Almonds, Apricots, Grapes, Papayas, Peaches, Citrus Fruits, Potatoes

USDA CROP PROGRESS REPORT

Percent US Corn Good to Excellent this Week

	5-yr avg.	2012	Last week	This week
Good				
Excellent				

Percent US Soybeans Good to Excellent this Week

	5-yr avg.	2012	Last week	This week
Good				
Excellent				

What's left?
US grain stocks stored on and off the farm (in thousands of bushels)

	Date	2010	2011	2012 (as of Sept 2012)	2013 (as of June 28, 2013)
Corn	Mar 1	7,693,787	6,523,228	6,023,356	
	Jun 1	4,310,071	3,669,838	3,148,540	
	Sep 1	1,707,787	1,127,645	988,403	
	Dec 1	10,056,769	9,647,466		
Soybeans	Mar 1	1,270,068	1,248,800	1,374,488	
	Jun 1	571,123	619,283	667,475	
	Sep 1	150,885	215,013	169,417	
	Dec 1	2,278,084	2,369,885		

Marketing Check List

❑ Review the USDA Acreage and USDA Grain Stocks reports.

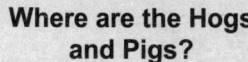

State	2012	As % of 2011
CO	720	99
IL	4,650	101
IN	3,800	101
IA	20,100	102
KS	1,790	96
MI	1,020	98
MN	7,800	100
MO	2,750	96
NE	3,100	98
NC	8,700	101
OH	2,120	103
OK	2,280	98
PA	1,110	100
SD	1,310	101
TX	870	130
UT	700	106
Oth Sts	3,009	96
US	**65,829**	**101**

Where are the Hogs and Pigs?
June 1 Inventory (1,000 head)

Inventory of all US Hogs & Pigs
(1,000 head)

Date of tally	2010	2011	2012	2013	2013 as % of 2011	2013 as % of 2012
Mar 1	63,568	63,684	64,937			
June 1	64,650	65,320	65,759			
Sept 1	65,971	67,234	67,472			
Dec 1	64,925	66,361				

July

7-13

New Crop This Week

	Hi	Lo	Close	12 Close
December Corn				$7.40
November Soybeans				$15.52

Old Crop This Week

	Hi	Lo	Close	12 Close
July Corn				$7.56
July Soybeans				$16.42

Sunday 7

Monday 8
- USDA report: Crop Progress
- LOW: Nov 2012 Soybeans, $8.60 on July 8, 2009

Tuesday 9
- Al's "Second Tuesday" Webinar

Wednesday 10

Thursday 11
- USDA reports: WASDE, Crop Production

Friday 12
- Last Trading Day: July Soybeans, July Corn

Saturday 13

My 2013 Beans: _____ acres X _____ bu/acre X $ _____ per bushel + $ _____
My 2013 Corn: _____ acres X _____ bu/acre X $ _____ per bushel + $ _____

	5-yr avg.	2012	Last week	This week
Percent US Corn Good to Excellent this Week				
Good				
Excellent				
Percent US Corn Silking this Week				
5-yr avg.		2012	Last week	This week
Percent US Soybeans Good to Excellent this Week				
Good				
Excellent				
5-yr avg.		2012	Last week	This week
Percent US Soybeans Setting Pods this Week				
5-yr avg.		2012	Last week	This week

T. B. or Not T. B. (Bureau of Animal Industry). 2 reels - 1,659 feet.

Fowl tuberculosis and methods of combating a plague that is taking a high toll among the flocks of Midwestern States.

¹ Available in both 35- and 16-millimeter widths.

USDA, *Division of Motion Pictures, 1934*

Al-ism

"The risk/reward ratio in farming hasn't changed much: Both are simply much higher than ever."

Marketing Check List

☐ Make sure you have all of your cash corn and soybeans sold by July 6.

#8. SPLIT FLOW & PHASING

Sometimes the jet stream will be coming into a particular region and it will split in two, sort of like when water in a stream or creek has to go around a large rock. In the atmosphere, the equivalent of a 'rock' could be a stalled weather system such as an Upper Level Low or a Dome.

The splitting of the Jet stream sets up the potential for a significant stormy pattern to develop IF the split in the jet stream reunites further to the east.

For example, we could see the jet stream splitting as it moves across the Pacific Ocean towards the West Coast of North America. Typically, one branch of the Jet stream goes up into Canada, then plunges down into the central and eastern US, bringing cold air. But the southern branch in the jet stream might be coming into Mexico or Texas, which could result in Low pressure forming over Texas and the Gulf coast.

When two or three jet streams merge into one big jet stream, that is called "phasing". It almost always results in a major weather system developing.

Al's Fearless Forecast

Al says: "Seasonal selling will work
in the grain markets in 2013."

January 1 US Cattle Inventory _(1,000 head)_					
	2011	**2012**	**2012 as % of 2011**	**July 19, 2013**	**2013 as % of 2012**
Cattle and Calves	100,000	97,800	98		
Cows and Heifers that have calved	40,600	39,700	98		
Heifers > 500 pounds	16,000	15,700	98		
Steers > 500 pounds	14,200	14,000	99		
Cattle on Feed	12,200	12,300	98		
Calf Crop (from prior year)	35,500	35,313.2	99		

5-Year Seasonal Odds this Week		
	change	reliability
Corn	-0.1¢ ⇩	80%
Soybeans	+10.0¢ ⇧	40%

New Crop This Week

	Hi	Lo	Close	12 Close
December Corn				$7.96
November Soybeans				$16.86

Old Crop This Week

	Hi	Lo	Close	12 Close
September Corn				$8.25
August Soybeans				$17.58

Sunday
14

Monday
15
- USDA report: Crop Progress

Tuesday
16
- Last Delivery Day: July Soybeans, July Corn

Wednesday
17

Thursday
18

Friday
19
- USDA report: Cattle On Feed, Cattle

Saturday
20

My 2013 Beans: _____ acres X _____ bu/acre X $ _____ per bushel + $ _____
My 2013 Corn: _____ acres X _____ bu/acre X $ _____ per bushel + $ _____

Percent US Corn Good to Excellent this Week	5-yr avg.	2012	Last week	This week
Good				
Excellent				

Percent US Corn Silking this Week	5-yr avg.	2012	Last week	This week

Percent US Soybeans Good to Excellent this Week	5-yr avg.	2012	Last week	This week
Good				
Excellent				

Percent US Soybeans Setting Pods this Week	5-yr avg.	2012	Last week	This week

Marketing Check List

☐ Watch corn crop conditions in July. This is the key month for determining corn yields.

21-27

5-Year Seasonal Odds this Week		
	change	reliability
Corn	-3.6¢ ⇩	60%
Soybeans	+1.6¢ ⇧	40%

New Crop This Week

	Hi	Lo	Close	12 Close
December Corn				$7.93
November Soybeans				$16.02

Old Crop This Week

	Hi	Lo	Close	12 Close
September Corn				$7.99
August Soybeans				$16.84

Sunday   21

Monday   22
- USDA report: Crop Progress

Tuesday   23

Wednesday   24

Thursday   25
- Al's "Last Thursday" Webinar

Friday   26

Saturday   27

My 2013 Beans: _____ acres X _____ bu/acre X $ _____ per bushel + $ _____
My 2013 Corn: _____ acres X _____ bu/acre X $ _____ per bushel + $ _____

28-3

5-Year Seasonal Odds this Week		
	change	reliability
Corn	-6.5¢ ⇩	40%
Soybeans	-30.0¢ ⇩	60%

New Crop This Week

	Hi	Lo	Close	12 Close
December Corn				$8.07
November Soybeans				$16.29

Old Crop This Week

	Hi	Lo	Close	12 Close
September Corn				$8.10
August Soybeans				$16.56

Sunday 28

Monday 29
- USDA report: Crop Progress

Tuesday 30

Wednesday 31

	Close Today	Close Last Month	Difference
Corn			
Soybeans			

Thursday 1

Friday 2

Saturday 3

My 2013 Beans: _____ acres X _____ bu/acre X $ _____ per bushel + $ _____
My 2013 Corn: _____ acres X _____ bu/acre X $ _____ per bushel + $ _____

Percent US Corn Good to Excellent this Week

	5-yr avg.	2012	Last week	This week
Excellent				
Good				

Percent US Corn Silking this Week

5-yr avg.	2012	Last week	This week

Percent US Soybeans Good to Excellent this Week

	5-yr avg.	2012	Last week	This week
Excellent				
Good				

Percent US Soybeans Setting Pods this Week

5-yr avg.	2012	Last week	This week

Al-ism

"People who do not like ethanol should not drive cars."

 Dutch Tip

"Spread The Joy": Pay your repair bills early (or at least on time) and you will be surprised at how fast your equipment gets fixed.

Marketing Check List

☐ Plan to attend a late-summer marketing seminar or Webinar.

5-Year Seasonal Odds this Week		
	change	reliability
Corn	+6.8¢ ⇧	60%
Soybeans	-4.0¢ ⇩	40%

New Crop This Week

	Hi	Lo	Close	12 Close
December Corn				$8.09
November Soybeans				$16.43

Old Crop This Week

	Hi	Lo	Close	12 Close
September Corn				$8.00
August Soybeans				$17.10

Sunday — 4

Monday — 5
- USDA report: Crop Progress

Tuesday — 6

Wednesday — 7

Thursday — 8

Friday — 9

Saturday — 10

My 2013 Beans: _____ acres X _____ bu/acre X $ _____ per bushel + $ _____
My 2013 Corn: _____ acres X _____ bu/acre X $ _____ per bushel + $ _____

In the USDA Crop Production Report this Month:

Barley, Corn, Hay, Oats, Rice, Sorghum, Wheat, Peanuts, Soybeans, Cotton, Sugarbeets, Sugarcane, Tobacco, Dry Edible Beans, Apples, Grapes, Olives, Papayas, Peaches, Pears, Prunes and Plums, Coffee, Ginger Root, Hops

USDA CROP PROGRESS REPORT

Percent US Corn Good to Excellent this Week	5-yr avg.	2012	Last week	This week
Good				
Excellent				

Percent US Corn Silking this Week				
	5-yr avg.	2012	Last week	This week

Percent US Soybeans Good to Excellent this Week	5-yr avg.	2012	Last week	This week
Good				
Excellent				

Percent US Soybeans Setting Pods this Week				
	5-yr avg.	2012	Last week	This week

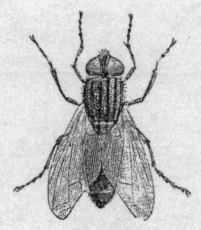

5-Year Seasonal Odds this Week		
	change	reliability
Corn	+15.0¢ ⇧	80%
Soybeans	-5.2¢ ⇩	60%

New Crop This Week

	Hi	Lo	Close	12 Close
December Corn				$8.07
November Soybeans				$16.46

Old Crop This Week

	Hi	Lo	Close	12 Close
September Corn				$7.99
September Soybeans				$16.71

Sunday — 11

Monday — 12
- USDA report: Crop Progress
- USDA reports: WASDE, Crop Production

Tuesday — 13
- Al's "Second Tuesday" Webinar

Wednesday — 14
- Last Trading Day: August Soybeans

Thursday — 15

Friday — 16
- Last Delivery Day: August Soybeans

Saturday — 17

My 2013 Beans: _____ acres X _____ bu/acre X $ _____ per bushel + $ _____
My 2013 Corn: _____ acres X _____ bu/acre X $ _____ per bushel + $ _____

The Cows Are Building Waterslides

The cows are building waterslides.
The pigs are playing pool.
A talking duck with rainbow stripes
is dancing on a stool.

The bulls are on the farmer's roof—
They're hiding from the goats.
I can't believe the baby chicks
are drinking root beer floats.

The hares are giving haircuts
to the tourists on the farm.
A crow just painted ice cubes
on the inside of my arm.

It's strange to see the peacocks
singing opera with the sheep.
I love the things my mind creates
when I am fast asleep!

Percent US Corn Good to Excellent this Week			
5-yr avg.	2012	Last week	This week
Good			
Excellent			

Percent US Corn Silking this Week			
5-yr avg.	2012	Last week	This week

Percent US Soybeans Good to Excellent this Week			
5-yr avg.	2012	Last week	This week
Good			
Excellent			

Percent US Soybeans Setting Pods this Week			
5-yr avg.	2012	Last week	This week

Marketing Check List

☐ Watch the trend in soybean crop conditions. This is a key month to determine soybean yields.

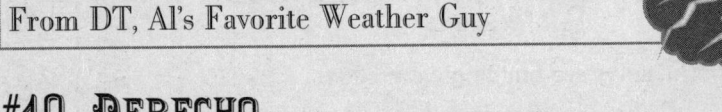

12 BIG WORDS O' WEATHER
From DT, Al's Favorite Weather Guy

#10. DERECHO

A derecho is a widespread severe wind event resulting from persistent and violent outflow from a MCS (see Word o' Weather #9). The derecho environment includes dry mid-levels winds that are part of a squall line or a segment of a squall line. The forward motion of the storm, along with an intense downdraft that occurs when there is heavy rain, causes a section of a line of strong thunderstorms to expand in a curve band as the Derecho moves east or southeast.

A severe wind is one with wind speeds of 50 knots (58 miles per hour) at the surface. In a derecho, these severe winds encompass a distance of at least 400 km (250 miles) either out ahead of, or along a squall line. The length of time the severe winds last can be particularly damaging. While a severe thunderstorm may produce severe convective wind gusts that last for several minutes at a point location, derecho wind can last 30 minutes or longer.

Derechos can be tracked from radar and severe weather reports while they are occurring, since severe weather reports will be given in sequence as the derecho traverses along.

In addition, most of these derecho have a fast forward speed that is equal to 50 or 60 MPH. This speed can sometimes add to the damaging winds from the actual system. However, this speed also restricts the ability of a derecho to produce significant tornado activity. Research shows that most of these derechos occur during a prolonged heat wave that features several days of temperatures in the 95-105° range.

18-24

5-Year Seasonal Odds this Week		
	change	reliability
Corn	+5.2¢ ⇧	40%
Soybeans	+52.0¢ ⇧	100%

New Crop This Week

	Hi	Lo	Close	12 Close
December Corn				$8.08
November Soybeans				$17.32

Old Crop This Week

	Hi	Lo	Close	12 Close
September Corn				$8.02
September Soybeans				$17.38

Sunday
18

Monday
19
- USDA report: Crop Progress

Tuesday
20

Wednesday
21
- HIGH: Dec 2012 Corn, $8.40 on Aug 21, 2012

Thursday
22

Friday
23
- USDA report: Cattle On Feed
- HIGH: Dec 2014 Corn, $6.16 on Aug 23, 2011

Saturday
24

My 2013 Beans: _____ acres X _____ bu/acre X $ _____ per bushel + $ _____
My 2013 Corn: _____ acres X _____ bu/acre X $ _____ per bushel + $ _____

	5-yr avg.	2012	Last week	This week
Percent US Corn Good to Excellent this Week				
Good				
Excellent				
Percent US Corn Silking this Week				
Percent US Soybeans Good to Excellent this Week				
Good				
Excellent				
Percent US Soybeans Setting Pods this Week				

I BUSHEL

Al's Fearless Forecast

Al says: "Brazil will produce more soybeans than the US in 2013."

Marketing Check List

☐ Livestock feeders and ethanol shareholders should use a late August break to start buying corn and meal through the end of 2012.

3 Scenarios For 2013

Here are three different scenarios that could unfold by the time your combines roll in the fall of 2013.

1. The bearish deflationary price scenario. In this scenario, the economic problems in Europe create a worldwide recession that throws China's economy into a major recession. Global stock and commodity markets move lower, and grain prices drop back to the levels of years 2000 to 2005. The chance of this scenario, I believe, is less than one in 10.

2. The inflationary price scenario. In this scenario, the gold and crude oil markets rally to new all-time highs, the grain markets follow along, and locking in inputs and land costs are major challenges. With the way governments around the world are printing money, I would give this scenario about one in 5.

3. The most likely scenario is for December 2013 corn to trade between $4.80 and $6.50 from now to October 2013. Get ready because the news will be really bearish when/if December 2013 corn futures drop below $5, just like it will be really bullish if futures go over $6.50.

One thing that is certain: It will be volatile. Keep in mind that in volatile years, farmers tend to make two types of marketing mistakes. One is getting too much sold ahead too early. The other is holding on too long.

How should you deal with the drama--both the giddy excitement, and the sinking fear--that comes with volatility? Here's how: Make a plan ahead of time, and stick with it. Know that volatility is here to stay, and work around it.

What's in Egypt's **CASE 30 stock index**?

Alexandria Mineral Oils
Amer Group Holding
Arabia Cotton Ginning
Arabia Investments Development
Cairo for Housing and Development
Citadel Capital
Commercial International Bank Egypt
Delta for Construction and Rebuilding
EFG Hermes Holding
Egypt Kuwait Holding
Egyptian Financial and Industrial
Egyptian Resorts
El Sewedy Electric
Ezz Steel
Juhayna Food Industries
Maridive and Oil Services
Modern Waterproofing
National Real Estate Bank for Development
National Societe Generale Bank
Orascom Construction Industries
Orascom Telecom Holding
Orascom Telecom Media and Technology Holding
Oriental Weavers Carpet
Palm Hills Developments
Pioneers Holding Co. for Financial Investment
Raya Holding Co. for Tech. and Telecommunication
Sidi Kerir Petrochemicals
Sixth of October Development and Investment
Talaat Mostafa Group Holding
Telecom Egypt

Egypt's economy--at least as represented by the CASE 30 stock index--was the top global performer. So what are those companies that are booming in Egypt? The CASE (Cairo & Alexandria Stock Exchange) 30, also called the EGX 30, is made up of the 30 most active stocks in the Egyptian stock market.

25-31

New Crop This Week

	12 Close	Close	Lo	Hi	
December Corn	$8.00				
November Soybeans	$17.56				

Sunday — 25

Monday — 26
- USDA report: Crop Progress

Tuesday — 27

Wednesday — 28

Old Crop This Week

	12 Close	Close	Lo	Hi	
September Corn	$8.03				
September Soybeans	$17.56				

Thursday — 29
- Al's "Last Thursday" Webinar

Friday — 30
- HIGH: Dec 2011 Corn, $7.77 on Aug 30, 2011

	Close Today	Close Last Month	Difference
Corn			
Soybeans			

Saturday — 31

My 2013 Beans: _____ acres X _____ bu/acre X $ _____ per bushel + $ _____
My 2013 Corn: _____ acres X _____ bu/acre X $ _____ per bushel + $ _____

Percent US Corn Good to Excellent this Week

	5-yr avg.	2012	Last week	This week
Good				
Excellent				

Percent US Soybeans Good to Excellent this Week

	5-yr avg.	2012	Last week	This week
Good				
Excellent				

Percent US Soybeans Setting Pods this Week

5-yr avg.	2012	Last week	This week

Al's Fearless Forecast

Al says: "Total corn revenue will exceed $90 billion in 2013."

Plows, Planes, and Peace (2 reels, 16 mm. and 35 mm., sound, released 1941).
Shows how America's farmers, through the ever-normal granary and other parts of the AAA program, have provided abundant food supplies to meet the demands for national defense.

USDA, *Division of Motion Pictures, 1941*

Marketing Check List

❏ Check your grain storage and augers to make sure everything is working.

Who's using up their leftover corn? (in millions of metric tons, as of Sept 2012)				
	Ending Stocks			
Country	**2009**	**2010**	**2011**	**2012**
China	51.3	49.4	59.4	60.2
US	43.4	28.6	30.0	18.6
Brazil	10.0	10.3	15.3	15.1
Europe (EU-27)	5.0	4.9	6.1	4.3
South Africa	5.2	3.4	2.7	2.7
FSU-12*	1.4	1.9	2.1	2.0
SE Asia	2.8	3.1	3.3	2.9
Mexico	1.4	1.4	0.9	1.2
Ukraine	0.7	1.1	1.2	1.3
Argentina	0.9	4.1	1.4	2.2
So. Korea	1.6	1.6	1.5	1.4
Canada	1.8	1.3	1.3	1.0
Egypt	1.5	1.3	1.8	1.1
Japan	0.7	0.6	0.6	0.6
WORLD	**143.9**	**127.58**	**139.6**	**124.0**

* "Former Soviet Union 12".

WHO ARE THE "FSU-12"?

The "Former Soviet Union Twelve" are 12 independent countries that were once part of the USSR:

1. Republic of Armenia
2. Republic of Azerbaijan
3. Republic of Belarus
4. Republic of Georgia
5. Republic of Kazakhstan
6. Republic of Kyrgyzstan
7. Republic of Moldova
8. Russian Federation
9. Republic of Tajikistan
10. Turkmenistan
11. Ukraine
12. Republic of Uzbekistan

5-Year Seasonal Odds this Week		
	change	reliability
Corn	+2.6¢ ⇧	60%
Soybeans	+55.0¢ ⇧	40%

New Crop This Week		Hi	Lo	Close	12 Close
	December Corn				$8.00
	November Soybeans				$17.36

Sunday
1

Monday
2
- CBOT Closed: Labor Day

Tuesday
3
- USDA report: Crop Progress

Wednesday
4
- LOW: Dec 2011 Corn, $3.71 on Sept 4, 2009
- LOW: Dec 2012 Corn, $3.87 on Sept 4, 2009
- HIGH: Nov 2012 Soybeans, $17.89 on Sept 4, 2012
- HIGH: Nov 2013 Soybeans, $14.09 on Sept 4, 2012

Thursday
5

Friday
6

Saturday
 7
- HIGH: Dec 2013 Corn, $6.65 on Sept 7, 2012

My 2013 Beans: _____ acres X _____ bu/acre X $ _____ per bushel + $ _____
My 2013 Corn: _____ acres X _____ bu/acre X $ _____ per bushel + $ _____

In the USDA Crop Production Report this Month:

Barley, Corn, Hay, Oats, Rice, Sorghum, Wheat, Peanuts, Soybeans, Cotton, Sugarbeets, Sugarcane, Tobacco, Dry Edible Beans, Apples, Grapes, Olives, Papayas, Peaches, Pears, Prunes and Plums, Coffee, Ginger Root, Hops

USDA CROP PROGRESS REPORT

Percent US Corn Good to Excellent this Week	5-yr avg.	2012	Last week	This week
Good				
Excellent				

Percent US Soybeans Good to Excellent this Week	5-yr avg.	2012	Last week	This week
Good				
Excellent				

Percent US Soybeans Setting Pods this Week	5-yr avg.	2012	Last week	This week

Al-ism

"Hot tips lead to burned fingers."

Dutch Tip

"Dutch Holiday": Farmers who participate in Webinars save travel time and money.

Marketing Check List

- ☐ This is usually one of the worst weeks of the year to sell cash soybeans because of low futures and wide basis levels.

September

8-14

New Crop This Week		12 Close	Close	Lo	Hi
	December Corn	$7.78			
	November Soybeans	$16.88			

Sunday
8

Monday
9
- USDA report: Crop Progress

Tuesday
10
- Al's "Second Tuesday" Webinar

Wednesday
11

Thursday
12
- USDA reports: WASDE, Crop Production

Friday
13
- Last Trading Day: Sept Soybeans, Sept Corn

Saturday
14
- HIGH: Nov 2014 Soybeans, $13.33 on Sept 14, 2012

My 2013 Beans: _____ acres X _____ bu/acre X $ _____ per bushel + $ _____
My 2013 Corn: _____ acres X _____ bu/acre X $ _____ per bushel + $ _____

Who had the most SOYBEANS left in their bins in 2012?
(in millions of metric tons, as of Sept 2012)

Country	Beginning Stock	Ending Stock
Argentina	18.47	20.12
Brazil	13.99	16.25
China	14.91	11.74
US	3.55	3.13
Europe (EU-27)	0.33	0.15
Japan	0.14	0.11
Mexico	0.07	0.02
WORLD	**53.65**	**53.10**

Who USES the most (and least) soybeans?
(in millions of metric tons, as of Sept 2012)

Country	2010	2011	2012
China	66.0	70.1	75.0
US	48.4	48.9	43.9
Argentina	39.2	37.6	39.9
Brazil	39.3	39.7	39.9
Europe (EU-27)	13.5	12.5	12.0
Japan	3.2	3.0	2.9
Mexico	3.7	3.6	3.5
WORLD	**251.8**	**254.5**	**256.7**

Percent US Corn Good to Excellent this Week

	Good	Excellent	5-yr avg.	2012	Last week	This week

Percent US Soybeans Good to Excellent this Week

	Good	Excellent	5-yr avg.	2012	Last week	This week

Marketing Check List

❑ If you have early varieties of soybeans harvested early, you may find some hungry processors.

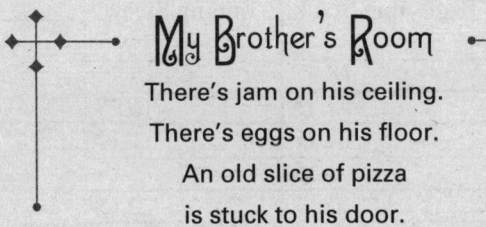

My Brother's Room

There's jam on his ceiling.
There's eggs on his floor.
An old slice of pizza
is stuck to his door.

His windows are filthy.
His dresser is dusty.
The bread on his bookshelf
is purple and crusty.

His tee shirts are scattered—
They cover his bed.
His hats are all stained
from the sweat on his head.

The worms in his closet
are slimy and big.
My big brother's bedroom
is fit for a pig!

US Corn Supply and Use
(millions of bushels)

Item	Crop Year			
	2010	2011	2012 Estimated (as of Sept 2012)	2013 Projected (as of Sep 12 2013)
Area planted (mil. ac.)	88.2	91.9	96.4	
Area harvested (mil. ac.)	81.4	84.0	87.4	
Yield per acre (bu)	152.8	147.2	122.8	
Beginning stocks	1,708	1,128	1,181	
Production	12,447	12,358	10,727	
Imports	28	25	75	
Total Supply	**14,182**	**13,511**	**11,983**	
Feed and residual	4,793	4,400	4,150	
Food, seed & industrial*	6,428	6,390	5,850	
Exports	1,834	1,540	1,250	
Total Use	**13,055**	**12,330**	**11,250**	
ENDING STOCKS	**1,128**	**1,181**	**733**	

* This includes:

Corn used for ethanol	5,021	5,000	4,500	

US Soybean Supply and Use
(in millions of bushels)

Item	Crop Year			
	2010	2011	2012 Estimated (as of Sept 2012)	2013 Projected (as of Sep 12, 2013)
Area planted (mil. ac.)	77.4	75.0	76.1	
Area harvested (mil. ac.)	76.6	73.6	74.6	
Yield per acre (bu)	43.5	41.5	35.3	
Beginning stocks	151	215	130	
Production	3,329	3,056	2,634	
Imports	14	16	20	
Total Supply	**3,495**	**3,287**	**2,785**	
Crushings	1,648	1,705	1,500	
Exports	1,501	1,360	1,055	
Seed	87	88	89	
Residual	44	3	25	
Total Use	**3,280**	**3,157**	**2,670**	
ENDING STOCKS	**215**	**130**	**115**	

15-21

5-Year Seasonal Odds this Week		
	change	reliability
Corn	-11.0¢ ⇩	40%
Soybeans	-30.0¢ ⇩	60%

New Crop This Week		Hi	Lo	Close	12 Close
	December Corn				$7.48
	November Soybeans				$16.22

Sunday 15

Monday 16
- USDA report: Crop Progress

Tuesday 17
- Last Delivery Day: Sept Soybeans, Sept Corn

Wednesday 18

Thursday 19

Friday 20
- USDA report: Cattle On Feed

Saturday 21

My 2013 Beans: _____ acres X _____ bu/acre X $ _____ per bushel + $ _____
My 2013 Corn: _____ acres X _____ bu/acre X $ _____ per bushel + $ _____

Percent US Corn Good to Excellent this Week

	5-yr avg.	2012	Last week	This week
Good				
Excellent				

Percent US Soybeans Good to Excellent this Week

	5-yr avg.	2012	Last week	This week
Good				
Excellent				

Marketing Check List

☐ In the next two to four weeks, livestock feeders and ethanol shareholder should get the balance of their 2013 corn and meal locked in, and at least 25-50% of the first six months of 2014.

5-Year Seasonal Odds this Week		
	change	reliability
Corn	-37.0¢ ⇩	80%
Soybeans	-76.0¢ ⇩	100%

New Crop This Week		Hi	Lo	Close	12 Close
	December Corn				$7.56
	November Soybeans				$16.01

Sunday
- First Day Of Fall

22

Monday
- USDA report: Crop Progress

23

Tuesday

24

Wednesday

25

Thursday
- Al's "Last Thursday" Webinar

26

Friday
- USDA report: Quarterly Hogs and Pigs

27

Saturday
- HIGH: Dec 2015 Corn, $6.00 on Sept 28, 2012

28

My 2013 Beans: _____ acres X _____ bu/acre X $ _____ per bushel + $ _____
My 2013 Corn: _____ acres X _____ bu/acre X $ _____ per bushel + $ _____

Percent US Corn Good to Excellent this Week

	5-yr avg.	2012	Last week	This week
Good				
Excellent				

Percent US Soybeans Good to Excellent this Week

	5-yr avg.	2012	Last week	This week
Good				
Excellent				

Inventory of all US Hogs & Pigs
(1,000 head)

Date of tally	2010	2011	2012	2013	2013 as % of 2011	2013 as % of 2012
Mar 1	63,568	63,684	64,937			
June 1	64,650	65,320	65,759			
Sept 1	65,971	67,234	67,472			
Dec 1	64,925	66,361				

Marketing Check List

❑ If prices have been trending lower, this is often the week that corn and soybean futures bottom.

Who USES the most (and least) wheat?
(in millions of metric tons, as of Sept 2012)

Country	2010-11	2011-12	2012-13
EU-27	122.0	126.5	124.5
China	110.5	120.5	122.0
FSU-12	75.04	80.2	74.00
India	81.8	81.5	86.9
Russia	38.6	38.0	35.5
North Africa	39.2	41.1	40.8
United States	**30.7**	**32.2**	**33.8**
Selected Mideast	33.0	34.0	32.1
Pakistan	23.0	23.1	23.2
Ukraine	11.6	15.0	12.3
SE Asia	14.5	16.4	16.1
Brazil	10.8	11.2	11.0
Canada	7.7	9.6	7.9
Kazakhstan	6.2	8.0	7.0
Australia	6.1	6.7	6.7
Argentina	6.0	6.0	6.0
WORLD	**654.5**	**694.4**	**680.66**

	Date	2010	2011	2012 (as of Sept 2012)	2013 (as of Sep 30, 2013)

What's left?

*US **grain stocks** stored on and off the farm (in thousands of bushels)*

	Date	2010	2011	2012 (as of Sept 2012)	2013 (as of Sep 30, 2013)
Corn	Mar 1	7,693,787	6,523,228	6,023,356	
	Jun 1	4,310,071	3,669,838	3,148,540	
	Sep 1	1,707,787	1,127,645	988,403	
	Dec 1	10,056,769	9,647,466		
Soybeans	Mar 1	1,270,068	1,248,800	1,374,488	
	Jun 1	571,123	619,283	667,475	
	Sep 1	150,885	215,013	169,417	
	Dec 1	2,278,084	2,369,885		

What I Watch

For the first 30 years of trading, I watched the U.S. global supply/demand reports, studied weather, and tried to anticipate fundamental changes.

All that is changed. I still stay aware of the fundamentals. But now when I write my morning email updates, this is what I look at, and in this order:

#1. Global stock markets. I watch the stock markets in China, Japan, Europe, and the U.S.

#2. Crude oil. The price of crude tends to pull corn prices higher or lower.

#3. U.S. dollar index. When the U.S. dollar is trending lower, commodities move higher and vice versa.

#4. Gold. When investors are buying gold, they are usually buying the rest of the commodity markets as well.

#5. Weather.

#6. Supply/demand.

What used to be my first factor is now my last – if I even write about it at all.

29-5

New Crop This Week		Hi	Lo	Close	12 Close
	December Corn				$7.48
	November Soybeans				$15.51

Sunday
29

Monday
30
- USDA report: Crop Progress
- USDA report: Grain Stocks

	Close Today	Close Last Month	Difference
Corn			
Soybeans			

Tuesday
1

Wednesday
2

Thursday
3

Friday
4

Saturday
 5

My 2013 Beans: _____ acres X _____ bu/acre X $ _____ per bushel + $ _____
My 2013 Corn: _____ acres X _____ bu/acre X $ _____ per bushel + $ _____

In the USDA Crop Production Report this Month:

Barley, Corn, Hay, Oats, Rice, Sorghum, Wheat, Peanuts, Soybeans, Cotton, Sugarbeets, Sugarcane, Tobacco, Dry Edible Beans, Apples, Grapes, Olives, Papayas, Peaches, Pears, Prunes and Plums, Coffee, Ginger Root, Hops

Percent US Corn Good to Excellent this Week	5-yr avg.	2012	Last week	This week
Good				
Excellent				

Percent US Soybeans Good to Excellent this Week	5-yr avg.	2012	Last week	This week
Good				
Excellent				

Al's Fearless Forecast

Al says: "Farmland will again outperform the US stock market."

Marketing Check List

❑ The quarterly USDA Grain Stocks report is one of the most important grain reports of the year. Plan to participate in our September 27 2013 Crop Report Webinar.

6-12

5-Year Seasonal Odds this Week		
	change	reliability
Corn	+17.0¢ ⇧	80%
Soybeans	+26.0¢ ⇧	60%

New Crop This Week		Hi	Lo	Close	12 Close
	December Corn				$7.53
	November Soybeans				$15.22

Sunday — 6

Monday — 7

Tuesday — 8
• Al's "Second Tuesday" Webinar

Wednesday — 9

Thursday — 10

Friday — 11
• USDA reports: WASDE, Crop Production
• HIGH: Nov 2015 Soybeans, $13.00 on Oct 11, 2012

Saturday — 12

My 2013 Beans: _____ acres X _____ bu/acre X $ _____ per bushel + $_____
My 2013 Corn: _____ acres X _____ bu/acre X $ _____ per bushel + $_____

Marketing Check List

❑ This is usually one of the worst weeks of the year to sell cash corn and soybeans, so it is also one of the best weeks to try and get cash corn and soybeans bought.

Your Long-Term Marketing Plan

How do you plan for the future? If you haven't gotten in this habit, let's start now. Ask yourself these five questions:

1. Do you have all of your cash corn and soybeans sold?

2. What percent of your 2013 crop do you have hedged ahead?

3. Do you have your inputs all locked in for 2013? How about 2014?

4. What percentage of your land do you own or have locked up with long-term leases?

5. Why are you worried today?

Your answers

1. Let's say you have most of your 2012 corn on hand, and half of your 2012 soybeans. First look at the current price level and do a risk-and-reward analysis. You have a lot more risk holding onto such a large part of your cash crop than you have in not yet doing any 2013 sales.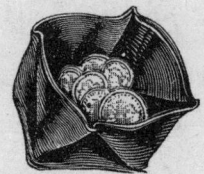

   What you can do: Make some cash sales today.

2. Let's say you have 30% of your new-crop 2013 corn hedged ahead, and about 20% of your new-crop 2013 soybeans contracted. These sales were put on in August 2012 at some very attractive price and profit levels.

   What you can do: Now ask yourself, "What is my plan for the next 10%?" If you don't have a marketing plan in writing yet, at least jot down some possible scenarios. Set them aside for a day, then look at them again. Which plan are you most comfortable with? Throw away the others.

3. Let's say you have all of your 2013 inputs bought, but you haven't even looked at 2014 inputs yet.

   What you can do: Look at the current prices of 2014 inputs. At least be aware of where they are compared to now.

4. Let's say you own half of the land you farm, and you have year-to-year cash-rent agreements on the other half.

   What you can do: How far in advance do you renew your cash-rent agreements? Can you get the landowner to renew a few months--or even a year--earlier?

5. Let's say you're nervous about 2013 because you read an article on the Internet forecasting cash corn to drop below $3 by the fall of 2013. You don't know who wrote it, or anything about the author's background. And really, what you're upset about is that you just lost a farm you have rented for 20 years, because you decided not to pay the increased rent.

What you can do: First, don't listen to Internet gossip from unreliable sources. Second, review your farm budget and marketing plan. Was it a good decision to give up the rented farm? If so, be at peace. You made a tough call, and will go on to make other good decisions. If you don't have a farm budget or marketing plan yet, put your anxiety to work, and write out the numbers. They will be your guide when you have to make the next tough call.

13-19

5-Year Seasonal Odds this Week		
	change	reliability
Corn	+2.1¢ ⇧	60%
Soybeans	-6.6¢ ⇩	40%

New Crop This Week		Hi	Lo	Close	12 Close
	December Corn				$7.61
	November Soybeans				$15.34

Sunday   13

Monday   14
- CBOT CLOSED: Columbus Day

Tuesday   15

Wednesday   16

Thursday   17

Friday   18
- USDA report: Cattle On Feed

Saturday   19

My 2013 Beans: _____ acres X _____ bu/acre X $ _____ per bushel + $ _____
My 2013 Corn: _____ acres X _____ bu/acre X $ _____ per bushel + $ _____

Al-ism

"Treat your bins like banks... before the thieves do."

PROGRESS IN MOTION PICTURES.

Marked advances made in the fiscal year 1920 in the production and distribution of the department's motion pictures leave no doubt as to the effectiveness and value of this mode of bringing directly to the people the knowledge developed by the department's investigations and of acquainting the public with the methods and significance of important lines of work being carried on by the department.

These forward steps may be described as follows:

Editorial and laboratory work was completed on 52 new motion-picture subjects, and they were added to the department's film library. The largest previous number of releases in a single fiscal year was 18, in the fiscal year 1919. A list of the new subjects follows:

Department Circular 114, "Motion Pictures of the U. S. Department of Agriculture," was issued, giving the first compact and complete summary of information needed by department field workers using motion pictures, and containing a complete list of films released up to the end of the fiscal year.

A survey of the motion-picture field in the State agricultural extension organizations was completed. This survey indicates an almost unanimous desire to use motion pictures in agricultural extension work, and leaves no doubt that the motion picture will be a really powerful aid in this field as soon as extension workers can more generally obtain projection machines, and as soon as the available subjects and positive prints can be made more nearly equal to the demand.

USDA, *Division of Motion Pictures, 1920*

Marketing Check List

 ☐ Cash in the balance of your new crop corn and soybean  puts. Check the carrying costs for November-to-January soybean futures, and December-to-March corn futures; this is a key week to get those hedges rolled ahead.

October

20-26

5-Year Seasonal Odds this Week		
	change	reliability
Corn	+0.2¢ ⇧	60%
Soybeans	+18.0¢ ⇧	80%

New Crop This Week

	Hi	Lo	Close	12 Close
December Corn				$7.38
November Soybeans				$15.61

Sunday — 20

Monday — 21

Tuesday — 22

Wednesday — 23

Thursday — 24

Friday — 25

Saturday — 26

My 2013 Beans: _____ acres X _____ bu/acre X $ _____ per bushel + $ _____
My 2013 Corn: _____ acres X _____ bu/acre X $ _____ per bushel + $ _____

FARM MANAGEMENT

Magic in It (Bureau of Agricultural Economics). 2 reels - 1,277 feet.

Points out the advantages of using a farm-accounting system that makes possible a yearly summary of the farm business. Motion-picture "magic" is used to emphasize the lesson taught. Of interest to all farmers.

How About a Combine? (Bureau of Agricultural Economics). 1 reel - 998 feet.

The combined harvester and thresher as compared with the binder; the possibilities of the combine in the Eastern States.

USDA, *Division of Motion Pictures, 1937*

Marketing Check List

❏ Watch for a low in soybean futures during the World Series.

12 Big Words o' Weather
From DT, Al's Favorite Weather Guy

#11. Soil moisture

Soil moisture is important to forecasting. It affects both the temperature forecast and precipitation forecast.

High soil moisture will produce high evaporation, especially if temperatures warm significantly during the day. This evaporation will produce evaporative cooling. Although the temperature warms during the day, the evaporation keeps the soil from getting as warm as it otherwise would have.

A good example: Mississippi and Arizona in the summer. Both states are on about the same latitude. Mississippi generally has a high soil moisture content and evapotranspiration; Arizona generally has dry soils and low values of evapotranspiration.

High temperatures in southern Arizona average over 100 degrees in the summer. However, temperatures generally do not rise above 100 in Mississippi unless a drought reduces soil moisture. The normal humidity does make it feel more uncomfortable outside, though. It is hotter in Arizona, but the humidity in Mississippi can make it feel just as hot in the summer. High soil moisture values will tend to increase the dewpoint.

This has a major consequence for forecasted lows. The overnight low under uniform weather conditions will not drop by more than a couple degrees below the evening dewpoint, especially if the dewpoint is above 60 F. Condensation (a warming process) occurs when the temperature tries to drop below the dewpoint at night. Therefore, high dewpoints limit the amount of overnight cooling. If dewpoint are low, such as when a continental high pressure is in place or a location is located in a dry climate (or dry weather pattern), the overnight low will be much cooler than the afternoon high. Since the dewpoint is low, the temperature can continue falling at night without condensation warming the air and limiting the cooling. Rule of thumb: If the dew point depression (that is, the difference between the temperature and the dewpoint) is large during the afternoon, there will be a large temperature range between the high and low temperature.

Soil moisture is also important to precipitation forecasts. High soil moisture increases the likelihood of moisture convergence. A trigger mechanism--such as a front or low pressure--will not produce thunderstorm precipitation unless there is moisture in place to lift.

Moist air rising has a much better chance of producing precipitation than dry air rising. High soil moisture continuously evaporates moisture into the air, which helps to supply low-level moisture. The best combination is to have moist soils along with moisture being advected from a moisture source such as the Gulf of Mexico into a trigger mechanism.

Droughts and super saturated soils can produce a positive feedback loop that can continue the drought or flood. When the soils dry out, there is less moisture for fronts and other trigger mechanisms to lift, and therefore there is a continuation of less rainfall. On the other hand, when the soils are very saturated, the supply of evaporating moisture to the atmosphere is continuous and there is always moisture in place for a trigger mechanism to lift. It takes a dramatic shift in the weather pattern sometimes to end a drought or flood because of this positive feedback loop.

27-2

5-Year Seasonal Odds this Week		
	change	reliability
Corn	-3.3¢ ⇩	20%
Soybeans	-7.0¢ ⇩	80%

New Crop This Week		Hi	Lo	Close	12 Close
	December Corn				$7.40
	November Soybeans				$15.27

Sunday 27

Monday 28

Tuesday 29

Wednesday 30

Thursday 31
- Halloween
- Al's "Last Thursday" Webinar

	Close Today	Close Last Month	Difference
Corn			
Soybeans			

Friday 1

Saturday 2

My 2013 Beans: _____ acres X _____ bu/acre X $ _____ per bushel + $ _____
My 2013 Corn: _____ acres X _____ bu/acre X $ _____ per bushel + $ _____

In the USDA Crop Production Report this Month:

Barley, Corn, Hay, Oats, Rice, Sorghum, Wheat, Peanuts, Soybeans, Cotton, Sugarbeets, Sugarcane, Tobacco, Dry Edible Beans, Apples, Grapes, Olives, Papayas, Peaches, Pears, Prunes and Plums, Coffee, Ginger Root, Hops

Al-ism

"Marketing knowledge is power... and a better night's sleep."

Save the Soil (2 reels, 16 mm. and 35 mm., sound, released 1932; revised 1940).

With no more virgin land to exploit, the United States is faced with the problem of conserving its soil to prevent an agricultural decline. Shows various ways of conserving soil fertility, including terracing and cover-cropping to prevent erosion; crop rotation; use of legumes to build up the nitrogen and humus content of the soil; control of waste caused by fire; use of manures, and commercial fertilizers when necessary.

USDA, *Division of Motion Pictures, 1941*

 **Marketing Check List**

☐ For livestock feeders and ethanol shareholders: Increase your corn and meal coverage to at least 100% through the first six months of 2014.

Soybeans for Beans:
2012 Production
Forecasted by State (as of Oct 2012)

State	Area Harvested (1,000 Ac)	Yield (Bu/Ac)	Production (1,000 Bushels)
AL	330	39	12,870
AR	3,150	39	122,850
DE	168	38	6,384
GA	205	33	6,765
IL	8,800	39	343,200
IN	5,140	41	210,740
IA	9,290	43	399,470
KS	3,750	22	82,500
KY	1,450	37	53,650
LA	1,110	44	48,840
MD	475	42	19,950
MI	1,990	39	77,610
MN	6,970	43	299,710
MS	1,960	41	80,360
MO	5,250	30	157,500
NE	4,950	41	202,950
NJ	93	38	3,534
NY	307	45	13,815
NC	1,540	35	53,900
ND	4,700	34	159,800
OH	4,580	43	196,940
OK	300	20	6,000
PA	520	45	23,400
SC	370	30	11,100
SD	4,650	28	130,200
TN	1,220	35	42,700
TX	105	29	3,045
VA	580	39	22,620
WI	1,700	39	66,300
Others	40	40	1,587
US	**75,693**	**38**	**2,860,290**

3-9

New Crop This Week				
	December Corn	November Soybeans		
	Hi	Lo	Close	12 Close

Sunday — 3
• Daylight Savings Time Ends

Monday — 4

Tuesday — 5
• Election Day

Wednesday — 6

Thursday — 7

Friday — 8
• USDA reports: WASDE, Crop Production

Saturday — 9

My 2013 Beans: _____ acres X _____ bu/acre X $ _____ per bushel + $ _____
My 2013 Corn: _____ acres X _____ bu/acre X $ _____ per bushel + $ _____

US Corn Supply and Use
(millions of bushels)

Item	Crop Year			
	2010	2011	2012 Estimated (as of Sept 2012)	2013 Projected (as of Nov 8, 2013)
Area planted (mil. ac.)	88.2	91.9	96.4	
Area harvested (mil. ac.)	81.4	84.0	87.4	
Yield per acre (bu)	152.8	147.2	122.8	
Beginning stocks	1,708	1,128	1,181	
Production	12,447	12,358	10,727	
Imports	28	25	75	
Total Supply	**14,182**	**13,511**	**11,983**	
Feed and residual	4,793	4,400	4,150	
Food, seed & industrial*	6,428	6,390	5,850	
Exports	1,834	1,540	1,250	
Total Use	**13,055**	**12,330**	**11,250**	
ENDING STOCKS	**1,128**	**1,181**	**733**	

* This includes:

Corn used for ethanol	5,021	5,000	4,500	

Marketing Check List

☐ Check the carrying costs for November-to-January soybean futures, and December-to-March corn futures; this is a key week to get those hedges rolled ahead.

10-16

5-Year Seasonal Odds this Week		
	change	reliability
Corn	-16.0¢ ⇩	60%
Soybeans	-8.0¢ ⇩	40%

New Crop This Week	December Corn	November Soybeans	Hi	Lo	Close	12 Close

Sunday 10

Monday 11
• Veteran's Day

Tuesday 12
• Al's "Second Tuesday" Webinar

Wednesday 13

Thursday 14
• Last Trading Day: Nov Soybeans

Friday 15

Saturday 16

My 2013 Beans: _____ acres X _____ bu/acre X $ _____ per bushel + $ _____
My 2013 Corn: _____ acres X _____ bu/acre X $ _____ per bushel + $ _____

	Crop Year			
US Soybean Supply and Use *(in millions of bushels)*				
Item	2010	2011	2012 Estimated (as of Sept 2012)	2013 Projected (as of Nov 8, 2013)
Area planted (mil. ac.)	77.4	75.0	76.1	
Area harvested (mil. ac.)	76.6	73.6	74.6	
Yield per acre (bu)	43.5	41.5	35.3	
Beginning stocks	151	215	130	
Production	3,329	3,056	2,634	
Imports	14	16	20	
Total Supply	**3,495**	**3,287**	**2,785**	
Crushings	1,648	1,705	1,500	
Exports	1,501	1,360	1,055	
Seed	87	88	89	
Residual	44	3	25	
Total Use	**3,280**	**3,157**	**2,670**	
ENDING STOCKS	**215**	**130**	**115**	

Dutch Tip

"Joy of Knowledge": Money spent to attend a marketing education course might increase your income, and is tax-deductible. How much better can it get?

Marketing Check List

❑ Get at least 50% of your spring fuel needs locked in. Cooler temperatures in November and December usually drive up the price of fuel, gasoline and diesel.

Corn: 2012 Production
(Forecast as of Oct 2012)

State	Area harvested (1,000 ac)	Yield (bu/ac)	Production (1,000 bu)
AL	270	100	27,000
AR	690	177	122,130
CA	180	190	34,200
CO	970	138	133,860
DE	177	115	20,355
GA	295	190	56,050
IL	12,400	98	1,215,200
IN	6,050	100	605,000
IA	13,700	140	1,918,000
KS	4,200	91	382,200
KY	1,540	68	104,720
LA	530	170	90,100
MD	425	115	48,875
MI	2,340	118	276,120
MN	8,250	168	1,386,000
MS	780	156	121,680
MO	3,350	75	251,250
NE	9,150	142	1,299,300
NJ	82	132	10,824
NY	650	130	84,500
NC	780	120	93,600
ND	3,390	115	389,850
OH	3,620	123	445,260
OK	320	115	36,800
PA	1,000	127	127,000
SC	310	122	37,820
SD	5,350	94	502,900
TN	970	89	86,330
TX	1,540	145	223,300
VT	350	95	33,250
WA	115	210	24,150
WI	3,450	127	438,150
Others	497	161	79,955
US	87,721	122	10,705,729

The Al Kluis Farmer's Almanac

12 BIG WORDS O' WEATHER
From DT, Al's Favorite Weather Guy

#12. CAN IT BE TOO COLD TO SNOW?

You may have heard the saying, "it is too cold to snow today." While that is not technically correct, it is sometimes valid.

There are three requirements for making snow. First, the atmosphere must be cold enough to allow snow to reach the surface. Second, the atmosphere around that area must be saturated with moisture. And third, there must be some sort of mechanism around (such as a low pressure area and a cold front) so the cold air mass is lifted enough to form clouds, resulting in precipitation.

There are days when there is not enough moisture in the lower atmosphere to produce snow crystals or snow flakes in sufficient quantity to fall to the earth as snow. On the other hand, there may be enough moisture, but no way to lift the moisture up to produce the clouds needed to turn it into snow.

However there is no absolute rule about being too cold to snow. There have been many examples of major blizzards in the Midwest and on the East Coast when the temperature might be 10 or 12° and the dew point might be zero or 2° and it starts snowing... and sometimes snowing heavily.

However in severely cold air Arctic air masses, where temperatures at the surface reach 10 degrees F, the moisture capacity of the air will be so low that likely not much snow can occur. Only at these extremely low temperatures is the phrase "it is too cold to snow" fairly valid.

5-Year Seasonal Odds this Week		
	change	reliability
Corn	+2.0¢ ⇧	60%
Soybeans	+1.0¢ ⇧	60%

New Crop This Week

	Hi	Lo	Close	12 Close
December Corn				
November Soybeans				

Sunday
- LOW: Nov 2014 Soybeans, $10.54 on Nov 17, 2010

17

Monday
- Last Delivery Day: Nov Soybeans

18

Tuesday

19

Wednesday

20

Thursday
- Al's "Last Thursday" Webinar

21

Friday
- USDA report: Cattle On Feed

22

Saturday

23

My 2013 Beans: _____ acres X _____ bu/acre X $ _____ per bushel + $ _____
My 2013 Corn: _____ acres X _____ bu/acre X $ _____ per bushel + $ _____

24-30

5-Year Seasonal Odds this Week		
	change	reliability
Corn	-12.0¢ ⇩	40%
Soybeans	-4.0¢ ⇩	40%

New Crop This Week

	Hi	Lo	Close	12 Close
December Corn				
November Soybeans				

Sunday
24

Monday
25
- LOW: Nov 2015 Soybeans, $11.23 on Nov 25, 2011

Tuesday
26

Wednesday
27

Thursday
28
- CBOT Closed: Thanksgiving

Friday
29
- CBOT Early Close (12 Noon): Thanksgiving

	Close Today	Close Last Month	Difference
Corn			
Soybeans			

Saturday
30

My 2013 Beans: _____ acres X _____ bu/acre X $ _____ per bushel + $ _____
My 2013 Corn: _____ acres X _____ bu/acre X $ _____ per bushel + $ _____

Al-ism

"Who's that global businessman checking his bank accounts?
Oh... that's you, checking your bins!"

Suppressing Foot-and-Mouth Disease (1 reel, 16 mm. and 35 mm., sound, released 1939).

Short history of the disease in America, with map showing areas where outbreaks have occurred. How the disease may be transmitted. Eradication measures responsible for preventing spread of the disease in the United States: Rigid quarantine regulations; slaughter of infected or exposed animals and burial or cremation; thorough cleaning of premises; disinfection of trucks and railroad cars; inspection of animals in infected areas; removal of quarantine only when the area is proved to be freed of the infection. Includes scenes taken in California during the outbreak of the foot-and-mouth disease in 1924.

USDA, *Division of Motion Pictures, 1941*

Marketing Check List

❏ The week of Thanksgiving, you should have all corn and soybeans in commercial storage sold.

5-Year Seasonal Odds this Week		
	change	reliability
Corn	+17.0¢ ⇧	80%
Soybeans	+1.0¢ ⇧	40%

New Crop This Week

	Hi	Lo	Close	12 Close
December Corn				
November Soybeans				

Sunday 1

Monday 2

Tuesday 3

Wednesday 4

Thursday 5
- LOW: Nov 2010 Soybeans, $8.12 on Dec 5, 2008

Friday 6

Saturday 7

My 2013 Beans: _____ acres X _____ bu/acre X $ _____ per bushel + $ _____
My 2013 Corn: _____ acres X _____ bu/acre X $ _____ per bushel + $ _____

In the USDA Crop Production Report this Month:

Barley, Corn, Hay, Oats, Rice, Sorghum, Wheat, Peanuts, Soybeans, Cotton, Sugarbeets, Sugarcane, Tobacco, Dry Edible Beans, Apples, Grapes, Olives, Papayas, Peaches, Pears, Prunes and Plums, Coffee, Ginger Root, Hops

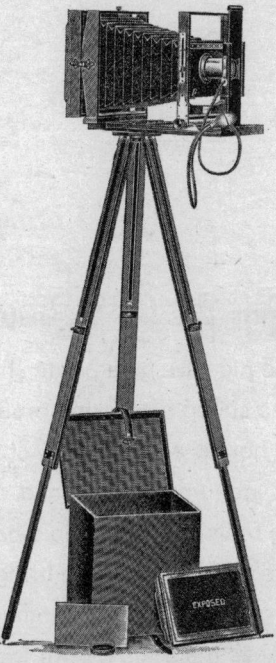

The River (3 reels, 16 mm. and 35 mm., sound, released 1939).

A dramatic documentary film of the Mississippi River—what it has done, and what man has done to it. A persuasive indictment of our practices of the past, and what we should do in the future if we are to avoid soil and lumber losses and the disastrous effects of floods. A conscious attempt to present a fundamental national problem so factually and so dramatically that those who see the picture will be moved to action.

USDA, Division of Motion Pictures, 1941

Marketing Check List

☐ Get copies of all your production records brought into your crop insurance agent.

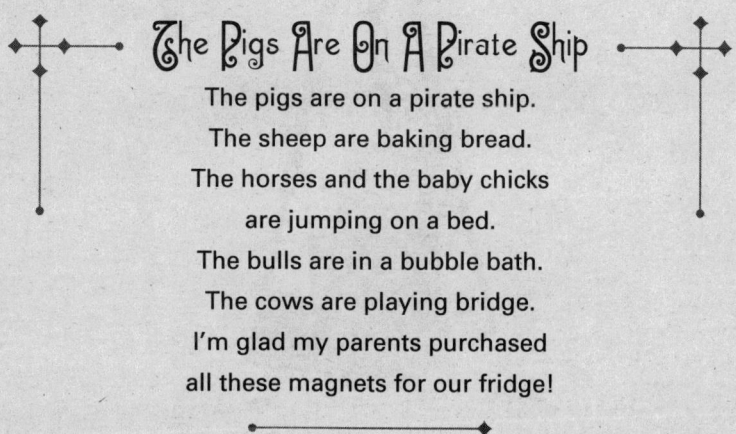

The Pigs Are On A Pirate Ship

The pigs are on a pirate ship.
The sheep are baking bread.
The horses and the baby chicks
are jumping on a bed.
The bulls are in a bubble bath.
The cows are playing bridge.
I'm glad my parents purchased
all these magnets for our fridge!

6-14

5-Year Seasonal Odds this Week		
	change	reliability
Corn	+13.0¢ ⇧	80%
Soybeans	+8.0¢ ⇧	80%

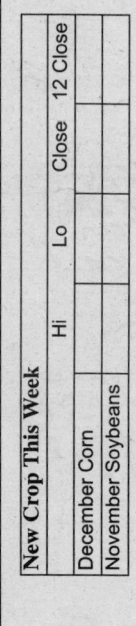

New Crop This Week

	Hi	Lo	Close	12 Close
December Corn				
November Soybeans				

Sunday 8

Monday 9

Tuesday 10
- USDA reports: WASDE, Crop Production
- Al's "Second Tuesday" Webinar

Wednesday 11

Thursday 12

Friday 13
- Last Trading Day: Dec Corn

Saturday 14

My 2013 Beans: _____ acres X _____ bu/acre X $ _____ per bushel + $ _____
My 2013 Corn: _____ acres X _____ bu/acre X $ _____ per bushel + $ _____

Al-ism

"Time you spend learning about marketing gets converted directly into money."

Bip Goes to Town (1 reel, 16 mm. and 35 mm., silent and sound, released 1941).

A small farm boy from an unelectrified dairy farm visits a modern electrified dairy farm and creamery and sees what electricity can do. He is excited at the prospect of electricity on his farm through an electrical cooperative.

USDA, Division of Motion Pictures, 1941

Marketing Check List

☐ Get your 2014 Al Kluis Farmer's Almanac ordered. Go to www.alkluis.com.

How the markets have changed since 2005

I have traded the grain markets for 37 years. The markets have changed dramatically since then, and how I now trade and how I approach the grain markets have also changed dramatically. For farmers, the ways you get your marketing information and the ways you should use that information have also changed. What caused these significant changes?

There were three major changes that happened between 2004 and 2007.

First, Wall Street firms started to view commodities as an asset class. After the Internet bubble and the drop in short-term and long-term interest rates, Wall Street companies, pension funds, and investors began looking at new ways to make money. They began taking a small part of their massive portfolios into the commodity markets. They put that money to work into commodity futures, commodity funds, commodity indexes, and commodity ETF's (exchange traded funds). This brought in new energy and a whole new dimension to the energy, grain, and livestock futures.

Second, the Chicago Board of Trade (CBOT) became part of the CME Group Inc. The Chicago Mercantile Exchange (CME) went public first and eventually merged with the CBOT. This changed how grain futures were traded and how customer accounts were handled. The focus of the CME was to drive trading volume by marketing to hedge funds, indexes, and Wall Street firms. Trading volume in the grain markets exploded.

Finally, the futures trade moved from the pit to electronic trading. This was part of the drive by the CME Group to make the markets more transparent, more liquid, and to expand the trading hours. Again the CME efforts really worked. Trading volume in the grain markets soared, and during a weather market in 2008, you could almost trade around the clock.

MOTION PICTURES

of the

UNITED STATES
DEPARTMENT *of*
AGRICULTURE

1934

MISCELLANEOUS PUBLICATION
NUMBER 208

15-21

5-Year Seasonal Odds this Week		
	change	reliability
Corn	+19.0¢ ⇧	100%
Soybeans	+31.0¢ ⇧	60%

New Crop This Week				12 Close
		Hi	Lo	Close
	December Corn			
	November Soybeans			

Sunday
15

Monday
16

Tuesday
17
- Last Delivery Day: Dec Corn

Wednesday
18

Thursday
19
- Al's "Last Thursday" Webinar

Friday
20
- USDA report: Cattle On Feed

Saturday
21
- First Day Of Winter

My 2013 Beans: _____ acres X _____ bu/acre X $ _____ per bushel + $_____
My 2013 Corn: _____ acres X _____ bu/acre X $ _____ per bushel + $_____

Behind the Breakfast Plate (Bureau of Animal Industry). 1 reel - 964 feet.

The story of bacon from the pastures of the Corn Belt through the stockyards and the packing plants to your breakfast plate; curing of bacon in early colonial homes; evolution of the hog business. Of general interest to consumers and producers.

USDA, *Division of Motion Pictures, 1934*

Marketing Check List

☐ Get your basis locked in on January Soybean hedges.

22-28

5-Year Seasonal Odds this Week		
	change	reliability
Corn	+10.0¢ ⇧	100%
Soybeans	+27.0¢ ⇧	80%

New Crop This Week			
	Hi	Lo	Close	12 Close
December Corn			
November Soybeans			

Sunday   22

Monday   23

Tuesday   24
- Christmas Eve

Wednesday   25
- CBOT Closed: Christmas Day

Thursday   26

Friday   27
- USDA report: Quarterly Hogs and Pigs

Saturday   28

My 2013 Beans: _____ acres X _____ bu/acre X $ _____ per bushel + $_____
My 2013 Corn: _____ acres X _____ bu/acre X $ _____ per bushel + $_____

Inventory of all US Hogs & Pigs
(1,000 head)

Date of tally	2010	2011	2012	2013	2013 as % of	
					2011	2012
Mar 1	63,568	63,684	64,937			
June 1	64,650	65,320	65,759			
Sept 1	65,971	67,234	67,472			
Dec 1	64,925	66,361				

Salt of the Earth (2 reels, 16 mm. and 35 mm., sound, released 1937). Portrays the farm family and the farm as the basis of our civilization. Shows how the national welfare and sustenance—bread, raiment, employment, wealth—depend on the remote activities of the farm. Under each of these headings a series of striking scenes of production dissolve into scenes illustrative of consumption or resultant urban industry. Each brings out the fundamental importance of the farm in the national economic scheme.

USDA, *Division of Motion Pictures, 1941*

Marketing Check List

☐ Make additional cash sales on any South American weather scare.

29-4

New Crop This Week		Hi	Lo	Close	12 Close
	December Corn				
	November Soybeans				

Sunday   29

Monday   30

Tuesday   31
• New Year's Eve

	Close Today	Close Last Month	Difference
Corn			
Soybeans			

Wednesday   1
• CBOT Closed: New Year's Day

Thursday   2

Friday   3

Saturday   4

My 2013 Beans: _____ acres X _____ bu/acre X $ _____ per bushel + $ _____
My 2013 Corn: _____ acres X _____ bu/acre X $ _____ per bushel + $ _____

Year-End Marketing Checklist.

- ❏ Select your marketing team and assign duties. You need a computer person, a chartist, and the person who keeps your records. Make sure you schedule monthly meetings.

- ❏ Plan on having someone from the team attend at least three marketing seminars or webinars this winter.

- ❏ Figure out the average selling price at which you sold your corn, soybeans, and wheat in 2013.

- ❏ List the highest sale prices you had for your corn, soybeans, and wheat.

- ❏ Look up what days your highest sales were made.

- ❏ List the lowest sale prices you had for your corn, soybeans, and wheat.

- ❏ Look up what days those sales were made.

- ❏ Write out the reason or reasons you made the sales when prices were high.

- ❏ Write out the reason or reasons you made the sales when prices were low.

- ❏ List three things you would like to change in the way you market your crops in 2014.

- ❏ Review the decision-making process you go through. Is it working? Why it should it change as market conditions change.

- ❏ Decide who you should have on your decision-making team in 2014. Set up your first monthly strategy meeting.

Marketing Check List

- ❏ Evaluate how you did in marketing in 2013. Look at who is on your marketing team and start thinking about who will be on your team in 2014.

2013

JANUARY
| | | | | | 1 | 2 | 3 | 4 | 5 |
6 7 8 9 10 11 12
13 14 15 16 17 18 19
20 21 22 23 24 25 26
27 28 29 30 31

FEBRUARY
1 2
3 4 5 6 7 8 9
10 11 12 13 14 15 16
17 18 19 20 21 22 23
24 25 26 27 28

MARCH
1 2
3 4 5 6 7 8 9
10 11 12 13 14 15 16
17 18 19 20 21 22 23
24 25 26 27 28 29 30
31

APRIL
1 2 3 4 5 6
7 8 9 10 11 12 13
14 15 16 17 18 19 20
21 22 23 24 25 26 27
28 29 30

MAY
1 2 3 4
5 6 7 8 9 10 11
12 13 14 15 16 17 18
19 20 21 22 23 24 25
26 27 28 29 30 31

JUNE
1
2 3 4 5 6 7 8
9 10 11 12 13 14 15
16 17 18 19 20 21 22
23 24 25 26 27 28 29
30

JULY
1 2 3 4 5 6
7 8 9 10 11 12 13
14 15 16 17 18 19 20
21 22 23 24 25 26 27
28 29 30 31

AUGUST
1 2 3
4 5 6 7 8 9 10
11 12 13 14 15 16 17
18 19 20 21 22 23 24
25 26 27 28 29 30 31

SEPTEMBER
1 2 3 4 5 6 7
8 9 10 11 12 13 14
15 16 17 18 19 20 21
22 23 24 25 26 27 28
29 30

OCTOBER
1 2 3 4 5
6 7 8 9 10 11 12
13 14 15 16 17 18 19
20 21 22 23 24 25 26
27 28 29 30 31

NOVEMBER
1 2
3 4 5 6 7 8 9
10 11 12 13 14 15 16
17 18 19 20 21 22 23
24 25 26 27 28 29 30

DECEMBER
1 2 3 4 5 6 7
8 9 10 11 12 13 14
15 16 17 18 19 20 21
22 23 24 25 26 27 28
29 30 31

2014

JANUARY

			1	2	3	4
5	6	7	8	9	10	11
12	13	14	15	16	17	18
19	20	21	22	23	24	25
26	27	28	29	30	31	

FEBRUARY

						1
2	3	4	5	6	7	8
9	10	11	12	13	14	15
16	17	18	19	20	21	22
23	24	25	26	27	28	

MARCH

						1
2	3	4	5	6	7	8
9	10	11	12	13	14	15
16	17	18	19	20	21	22
23	24	25	26	27	28	29
30	31					

APRIL

		1	2	3	4	5
6	7	8	9	10	11	12
13	14	15	16	17	18	19
20	21	22	23	24	25	26
27	28	29	30			

MAY

				1	2	3
4	5	6	7	8	9	10
11	12	13	14	15	16	17
18	19	20	21	22	23	24
25	26	27	28	29	30	31

JUNE

1	2	3	4	5	6	7
8	9	10	11	12	13	14
15	16	17	18	19	20	21
22	23	24	25	26	27	28
29	30					

JULY

		1	2	3	4	5
6	7	8	9	10	11	12
13	14	15	16	17	18	19
20	21	22	23	24	25	26
27	28	29	30	31		

AUGUST

					1	2
3	4	5	6	7	8	9
10	11	12	13	14	15	16
17	18	19	20	21	22	23
24	25	26	27	28	29	30
31						

SEPTEMBER

	1	2	3	4	5	6
7	8	9	10	11	12	13
14	15	16	17	18	19	20
21	22	23	24	25	26	27
28	29	30				

OCTOBER

			1	2	3	4
5	6	7	8	9	10	11
12	13	14	15	16	17	18
19	20	21	22	23	24	25
26	27	28	29	30	31	

NOVEMBER

						1
2	3	4	5	6	7	8
9	10	11	12	13	14	15
16	17	18	19	20	21	22
23	24	25	26	27	28	29
30						

DECEMBER

	1	2	3	4	5	6
7	8	9	10	11	12	13
14	15	16	17	18	19	20
21	22	23	24	25	26	27
28	29	30	31			

Questions for Our Curious Readers

1. What price did corn peak at in 2012?

2. What day did corn prices peak in 2012?

3. What day did soybean prices peak at in 2012?

4. Who produces more soybeans: Iowa or Argentina?

5. Who was the number two corn producing country in the world?

6. Who is the world's largest wheat producing country?

7. What percentage of the US corn crop is used for feed?

8. How much of this year's US soybean crop will be exported?

9. Where does the US rank in world wheat production?

10. Where does your state rank in corn and soybean production?

11. Which state has the most cattle in the US?

12. What was the total value of the US corn crop in 2012?

13. Who are the top two hog producing states in the US?

14. What percentage of the US corn crop is used for ethanol?

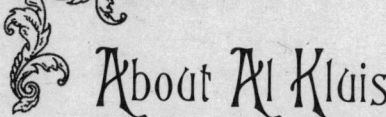

About Al Kluis

Al Kluis has been a commodity advisor and broker since 1976. He is president and managing partner of Kluis Commodities in Wayzata, Minnesota. Al is an introducing broker with R.J. O'Brien and provides marketing information and advice to farmers in 22 states and six countries.

Al is also a writer. His column, Your Profit, appears in every issue of *Successful Farming* magazine. Before that he was a contributing editor at *Corn and Soybean Digest* for 13 years; the magazine featured his Marketing Strategies column in every issue. Al has had two books on commodities trading published (his co-author on the first book, Loren Kruse, became Editor-in-Chief of Successful Farming magazine) and is commonly quoted in major publications including the *Wall Street Journal*. He is a featured speaker at commodity conferences nationwide and is a frequent market analyst for the Linder Farm Radio News Network. Al is the author of The Al Kluis Report.

Al, a Minnesota farmboy, was awarded his degree in Ag Economics from the University of Minnesota in 1974, after which he was executive director of the Minnesota Soybean Association before entering the markets full-time.

About Bill Lee

Bill Lee was born in Brooklyn, New York. He has won four international awards for cartoon artwork in magazines, was among the first cartoonists contacted for the newly-created *New York Times* op-ed page, and was one of the first journalists allowed into China at the opening of that country. Bill has had gallery exhibitions of paintings and sculptures and has done a screenplay for CBS, plus seven collections of cartoon art in humor books. He has traveled--writing and drawing--to China, Russia, Poland, Cuba, and beyond. Bill has covered three political conventions and gives lectures nd slideshows on humor in cartoon art. Bill has a daughter, Jennifer Catherine, and his hobbies are relaxing with a drink and music that ranges somewhere in between Hank Williams and Mozart.

About Darren Sardelli

Darren Sardelli is an award-winning poet and children's book author. Darren's poems are featured in 12 books in the US and the UK, as well as various other publications including school year books. Aside from writing, Darren enjoys bungee jumping, cliff diving, snorkeling, jet skiing, and building snowmen. He is a huge fan of ice hockey, and has played hockey in Germany, France, England, Ireland, Italy, Switzerland, and Canada. Darren lives on Long Island, Long Beach, New York. During the school year, Darren visits schools all over the country. His assemblies and workshops have inspired thousands of students to start writing funny poetry. Read Darren's new book,"Galaxy Pizza and Meteor Pie". Learn more about Darren at www.LaughALotPoetry. com.

About Katie Thompson

Katie Thompson is a writer and photographer who has published three books and hundreds of stories, columns, and photos in periodicals including *Successful Farming* magazine, *New Ag International*, the *Sioux City Journal*, and the *Des Moines Register*. She has also written patents (4 awarded), business plans, and questions for a quiz show called *"Smart Farmer."* Katie produces the annual *Kluis Commodity Calendar* and the *Al Kluis Farmer's Almanac*.

About David Tolleris

David Tolleris ("DT" to his friends and clients) is an energy and agricultural meteorologist based in Richmond, Virginia. DT graduated from City College of New York in 1989 and worked for several private weather forecasting firms in New York City and Hartford, Connecticut. In 1994 he joined the National Weather Service. In 1998, DT left the Federal Government to take care of his new son and to start his own weather business. Since then, DT has been forecasting for energy and agricultural companies. DT's website, www.wx-risk.com, is a very popular and well-known source of weather information for farmers, energy traders, and ag traders all around the world. "It's no-nonsense, no-BS information," says DT. He issues numerous reports daily, weekly and monthly via his website, with a special emphasis on 30-day forecasts and overseas ag concerns.

DT deliberately does not provide ANY trading advice. Instead, he focuses exclusively on his weather forecasts. "Which, believe me," he says, "is plenty hard enough to do." DT is a US Navy Veteran, a baseball fan, and a diehard, fanatical supporter of the Philadelphia Phillies.

If you liked this Handy Almanac, you'll love the Kluis Commodity Calendar and The Al Kluis Report.

The Kluis Commodity Calendar

In This Beautiful Oversized Calendar You Will Find:
- Seasonal Odds
- Monthly Marketing Checklists
- Thought of The Month
- News To Watch
- A Place To Track And Chart Your Weekly Basis

... And Beautiful Photography and Images You Will Find Nowhere Else, Illustrating The Places And Highlights Of The Business Of Agriculture.

The Al Kluis Report & The Al Kluis Market Pack

Every Saturday Morning, The Al Kluis Report Will Tell You:
- What To Watch
- What To Plan
- What To Do

Inside You Will Find:

- **Consistent Market Plans Like This:**
  "...Protect your 2013 crop income against a crop failure or sharply lower prices by implementing this three-part plan: #1. Get hedges on 30-50% of your 2013 crop corn, soybeans, and..."

- **Heads-up On News That Can Move Your Markets, Like This:**
  "...The farm strike in Argentina continues to be a problem...."

- **And Specific Marketing Recommendations, Like This:**
  "... Cash in your short $12.90 November Soybean futures and the $11.00 puts..."

  "...We recommend buying at least 50% of your fall dryer gas and winter propane early this week..."

  "...The basis should be set on all May - July hedge and hedge-to-arrive contracts..."

The Al Kluis Report

is only available as part of the comprehensive *Al Kluis Market Pack*, because if you're serious about marketing your grain well, you need the Al Kluis Market Pack.

The Al Kluis Market Pack Includes:

- **The Al Kluis Report,** every Saturday, 48 weeks per year (delivered promptly by e-mail or viewable at AlKluis.com)
- **Twice-Daily Marketing Updates** (accessible at a toll-free number, or viewable at AlKluis.com. AM Update delivered by email.)
- **CBOT Price Quotes by Text Message,** 3 or 5 times per market day
- **Action Alerts by Text Message,** one to three times per month
- **Attendance For 2** at two Market Outlook Seminars and all subscriberonly Webinars
- **Selected Special Reports from Al Kluis**
- **One Kluis Commodity Calendar**
- **Two Al Kluis Farmer's Almanacs** (one for your office, one for your truck)

The price for this package is $880 per year. Here are several ways to order:
- **Call 1-888-345-CORN (2676)**
- or Use the Order Form in the back of this Almanac
- or Go to www.AlKluis.com

You can pay by credit card, check, money order, or be invoiced. If you are serious about marketing your grain well, you need the *Al Kluis Market Pack*.